Biometric Quality
The last 1%: Biometric Quality Assessment for Error Suppression

NIST Interagency Report 7544

Elham Tabassi
Patrick Grother

Information Access Division

November 2008

Contents

1 What is Meant by Quality?   **11**
    1.1 Aspects of Quality 12
    1.2 Scalar vs. Vector Quality 12
    1.3 Number of Quality Values 15

2 Uses of Quality Assessments   **17**
    2.1 Quality at the Time of Capture 18
       2.1.1 Enrollment Phase Quality Assessment 18
       2.1.2 Verification Phase Quality Assessment 19
       2.1.3 Identification Quality Assessment 20
    2.2 Performance-Related Quality Monitoring 20
    2.3 Differential Processing 21

3 Standardization   **23**
    3.1 The ISO/IEC 29794 Biometric Sample Quality Standard 24
    3.2 The ANSI/NIST ITL 1-2007 Quality Field 25
    3.3 The BioAPI Quality Specification 27

4 Properties of a Quality Measure   **29**
    4.1 Notation 29
    4.2 Relationship to Matching 30
    4.3 Relationship to Performance 31
    4.4 Combining Quality Values 31

5	**Do Quality Values Predict Performance?**	**35**
	5.1 Rank-Ordered Detection Error Trade-off Characteristics	36
	5.2 Error vs. Reject Curves	39
	5.3 Generalization to Multiple Matchers	41
	5.4 Measuring Separation of Genuine and Impostor Distributions	43
	5.5 Data to be Used for Testing	43
6	**Quality Reference Data Sets**	**49**
	6.1 Construction of a Reference Data Set	49
	6.1.1 Data	49
	6.1.2 Target Quality Assignment	50
7	**Fingerprint Image Quality**	**53**
	7.1 NIST Fingerprint Image Quality (NFIQ)	53
	7.2 Recommendations for NFIQ Summarization	54
A	**Determination of Quality Weights**	**59**

List of Figures

1.1 Examples of (a) a low-character fingerprint due to poor skin condition, (b) a poor-quality facial image due to poor user behavior (non-frontal pose), (c) a poor-quality fingerprint due to unclean platen with residual fingerprint on it, and (d) a low-quality (out of focus) facial image because of imperfect acquisition setting. 13

2.1 Example of same-session captures of single finger that, despite their poor quality (NFIQ =5), were matched correctly by three leading commercial matchers. 19

3.1 Components of quality as defined by ISO/IEC 29794 Biometric Sample Quality - Part 1: Framework. The character of a sample indicates the richness of features from which the biometric sample is derived. The fidelity of a sample is the degree of similarity between a biometric sample and its source. The utility of a sample is indicative of positive or negative contribution of an individual sample to the overall performance of a biometric system. Source: ISO/IEC JTC1 SC 37 N2727. 26

3.2 Structure of header in a biometric data block as defined in ISO/IEC 19794-x. 26

4.1 Dependence of raw genuine scores on the transformed NFIQ qualities of the two input samples. 32

4.2 Boxplots of genuine scores, FNMR , impostor scores, and FMR for each of five transformed NFIQ quality levels for scores from a commercial matcher. Each quality bin, q, contains scores from comparisons of enrollment images with quality $q^{(1)} \geq q$ and subsequent use-phase images with $q^{(2)} = q$, per the discussion in Section ??. The boxplot notch shows the median, the box shows the interquartile range, and the whiskers show the extreme values. Notches in (d) are not visible, because the medians of FMR s are zero and therefore outside the plot range. 33

4.3 Dependence of the error vs. reject characteristic on the quality combination function H(.). (a) shows, for a fixed threshold, the decrease in FNMR as samples with low-quality scores are rejected. (b) shows the decrease in FMR at a fixed threshold, as samples with low-quality scores are rejected. Index fingerprints collected at operational settings were used. The similarity scores come from commercial matchers. 34

5.1 Quality-ranked detection error trade-off characteristics. Each plot shows five traces corresponding to five transformed NFIQ levels. 38

5.2 Error vs. reject performance for three fingerprint quality methods. Figures (a) and (b) show reduction in FNMR and FMR at a fixed threshold as up to 20 % of the low-quality samples are rejected. The similarity scores come from a commercial matcher. 40

5.3 Error vs. reject characteristics showing how NFIQ generalizes across (a) five verification algorithms, and (b) three operational data sets. The steps in (a) occur at the same rejection values because the matchers were run on a common database. 42

5.4 There is a higher degree of separation between the genuine and impostor distribution for better-quality samples as measured by NFIQ 44

5.5 Scatter plots of scores and FNMR values versus quality, and the error vs. reject curves, for a face quality metric applied to a face database composed of images at full (blue), half (green), and quarter size (red). 46

5.6 Error vs. reject characteristic of same algorithm as in Figure ?? but using data with less quality variation. The plots correspond to different quality combination function H(.) as discussed in Section ??. 47

6.1 Empirical cumulative distribution functions for the top-ranked genuine scores and for the impostor scores. The vertical lines are one possible way of binning normalized match scores. Samples are assigned quality numbers corresponding to the bin of their normalized match score. 52

7.1 Quality-ranked detection error trade-off characteristics. Five traces correspond to five NFIQ levels. Fingerprint images with NFIQ=1 (highest quality) cause lower recognition error than images with NFIQ=5 (lowest quality). 55

7.2 Dependance of NFIQ weights on operating threshold. Weights for NFIQ values 1 and 2 are quite robust to variation of the computing threshold. Thresholds are set at overall false-match-rates of 0.01, 0.001, and 0.0001. Each point corresponds to the NFIQ weight estimated using similarity scores of a commercial matching algorithm on large operational fingerprint datasets. NFIQ weights in Table ?? are means of six matching algorithms with the highest performance. 57

List of Tables

2.1 NFIQ summary of a random selection of fingerprint images collect as part of the US-VISIT program at two different locations over one week. 22

3.1 Structure of five-byte quality field that SC 37 Working Group 3 is considering. 27

3.2 BioAPI quality categories 28

5.1 KS statistics for quality levels of three quality algorithms 43

6.1 Binning normalized match score 52

7.1 Recommendation for NFIQ summarization at different operating thresholds 57

Executive Summary

Results from as early as Fingerprint Vendor Test 2003 clearly demonstrate that one of the most significant factors affecting biometric accuracy is that of quality. Test and evaluations demonstrate time and again that many algorithms perform well on high quality biometric samples, but what set them apart is how algorithms perform on poor quality samples. Although only a small fraction of input data are of poor quality, the bulk of recognition errors can be attributed to poor quality samples. Poor quality samples decrease the likelihood of a correct verification and/or identification, while extremely poor-quality samples might be impossible to verify and/or identify.

Biometric quality analysis is a technical challenge because it is most helpful when the measures reflect the performance sensitivities of one or more target biometric matchers. NIST addressed this problem in August 2004 when it issued NIST Fingerprint Image Quality (NFIQ) algorithm. NFIQ is a fingerprint quality measurement tool; it is implemented as open-source software; and is used today in U.S. government and commercial deployments. Its key innovation is to produce a quality value from a fingerprint image that is directly predictive of expected matching performance, and has been designed to be matcher independent. There is now international consensus in industry, academia, and government that a statement of a biometric sample's quality should be related to its recognition performance.

If quality can be improved, either by sensor design, by user interface design, or by standards compliance, better performance can be realized. For those aspects of quality that cannot be designed-in, an ability to analyze the quality of a live sample is needed. This is useful primarily in initiating the reacquisition from a user, but also for the real-time selection of the best sample, and the selective invocation of different processing methods. Accordingly, quality measurement algorithms are increasingly deployed in operational biometric systems. U.S. Visitor and Immigrant Status Indicator Technology (US-VISIT), U.S. governments Personal Identity Verication (PIV) program, the U.S. Department of Homeland Security's Transportation Worker Identication Credential (TWIC), and EU Visa Information System (VIS) each mandate the measurement and reporting of quality scores of captured images. With the increase in deployment of quality algorithms, the need to standardize an interoperable way to store and exchange biometric quality scores increases.

Recognizing this need, the Department of Homeland Securitys Science and Technology Directorate initiated a program with the National Institute of Standards and Technology to develop:

▷ open source software to compute quality score of biometric samples,
▷ tools and guidance on the wider use of quality measures in biometric systems including but not limited to quality summarization, examining methods of assessing how effective a quality algorithm is in predicting performance, and role of quality measures in multimodal biometric systems, and

▷ international standard that establishes an interoperable way of storing and exchanging biometric quality scores.

This document describes NIST's activities on biometric sample quality research and standardization. The main points of this document are summarized below.

Quality measurement plays vital role in improving biometric system accuracy and efficiency during the capture process (as a control-loop variable to initiate reacquisition), in database maintenance (sample update), in enterprise-wide quality assurance surveying, and in invocation of quality-directed processing of samples. Neglecting quality measurement will adversely impact accuracy and efficiency of biometric recognition systems (e.g., verification and identification of individuals).
Chapter 2

Biometric Quality Assessment (BQAM) algorithms shall produce quality scores that predict performance metrics such as either false match or false non-match. Thus, quality scores should reflect the sensitivities and failure modes of the matching algorithm. The term quality should not be solely attributable to the acquisition settings of the sample, such as image resolution, dimensions in pixels, grayscale/color bit depth, or number of features. Though such factors may affect sample utility and could contribute to the overall quality score.
Chapter 1

In January 2006, the Biometrics Subcommittee (SC 37) of Joint Technical Committee (JTC 1) initiated work on ISO/IEC 29794, a multipart standard that establishes quality requirements for generic aspects (Part 1), fingerprint image (Part 4), facial image (Part 5), and, possibly, other biometrics later. Part 1 of multipart ISO/IEC 29794 draft standard requires quality scores to be predictive of performance metrics such as either false match or false non-match, and defines a binary record structure for the storage of a sample's quality data.
Chapter 3

Recommend the use error vs. reject curves as a mean of evaluating (BQAM)s. The goal is to state how efficiently rejection of low-quality samples results in improved performance. This models the operational case in which quality is maintained by reacquisition after a low-quality sample is detected.
Chapter 5

Recommend a procedure to annotate the samples of a reference corpus with quality values. Quality-annotated corpus could be used for quality algorithm development, quality calibration, and conformity of quality scores to a standard.
Chapter 6

Review NIST Fingerprint Finger Image Quality and recommend procedures for NFIQ summarization. The motivation for NFIQ summarization is to monitor quality variation over time, across different acquisition settings and/or application.
Chapter 7

Chapter 1

What is Meant by Quality?

Broadly, a sample should be of good quality if it is suitable for automated matching. This viewpoint may be distinct from the human conception of quality. If, for example, an observer sees a fingerprint with clear ridges, low noise, and good contrast, then he might reasonably say it is of good quality. However, if the image contains few minutiae, then a minutiae-based matcher would underperform. Likewise, if a human judges a face image to be sharp, but a face recognition algorithm benefits from slight blurring of the image, then the human statement of quality is inappropriate. Thus the term quality should not be used to refer to the fidelity of the sample, but instead to the utility of the sample to an automated system. The assertion that performance is ultimately the most relevant goal of a biometric system implies that a Biometric Quality Assessment Method (BQAM) should reflect the sensitivities and modes of the matching algorithm. For fingerprint minutiae algorithms, this could be the ease with which minutiae are detected. For face algorithms, it might include how readily the eyes are located. The definition of quality as prediction of performance was first introduced by NIST when the agency released the NIST Fingerprint Image Quality (NFIQ) reference in August 2004 [7,8]. There is now international consensus in industry [3], academia [4], and government [5] that a statement of a biometric sample's quality should be related to its recognition performance. That is, a quality measurement algorithm takes a signal or image, $\mathbf{x}$, and produces a scalar, $q = Q(\mathbf{x})$, which is related monotonically to the performance of biometric matchers, under the constraint that at least two samples with their own qualities (as opposed to a pairwise quality) are being compared. A quality measure could be tuned to predict the performance of one matcher (the more common and useful case) or the more difficult case of one that generalizes to other matchers or classes of matchers.

1.1 Aspects of Quality

Source of quality impairments can be classified in the following four categories:

- **Character** The character of an image indicates the richness of identifying features and attributes affecting image processing, for example, extreme pupil dilation for iris recognition;
- **Behavior** The behavior of an image is a measure of optimality of user interaction with the capture device, for example, positioning, forcing of failures, etc.;
- **Imaging** The character of an image indicates image properties and characteristics, for example, resolution, focus, compression, distortion, etc.; and
- **Environment** The character of an image quality indicating aspects attributed to the acquisition settings, for example, illumination, background, etc.

Clearly many factor could affect each of the above quality aspects. Figure 1.1 shows images with specific quality impairments.

Taking proper action to improve quality of an acquired sample is possible only if the source of quality impairment is known. If a biometric sample is of low character (e.g., scar on a fingertip), recapture will not improve quality, but perhaps invoking different or extra processing methods (e.g., image enhancement or fusion) would improve the utility of the sample. If the cause of quality imperfection is subject behavior (e.g., a non-frontal facial image), a recapture with appropriate feedback to the subject (to directly look into the camera, for example) will result in acquisition of a better quality sample.

A scalar quality score reflects the predicted positive or negative contribution of an individual sample to the overall performance of a biometric system. However, it does not provide fine grade knowledge of the likely causes of quality imperfection. To make quality "actionable" by providing needed feedback to users, operators, or processing algorithms, an ability to track the source of quality imperfection to the four aspects of quality (character, behavior, imaging and environment) is desired. However, currently it is an extremely challenging problem, at least because of the following two reasons: a) sensitivity and behavior of capture devices and matching algorithms to the above mentioned quality aspects are not fully known and b) lack of proper data (samples with single specific defect) makes devising and implementing metrics to assess quality components of each aspect very difficult.

1.2 Scalar vs. Vector Quality

We have thus far suggested that biometric quality scores are scalars, as opposed to vectors, for example. Operationally the requirement for a scalar is not necessary: a vector could

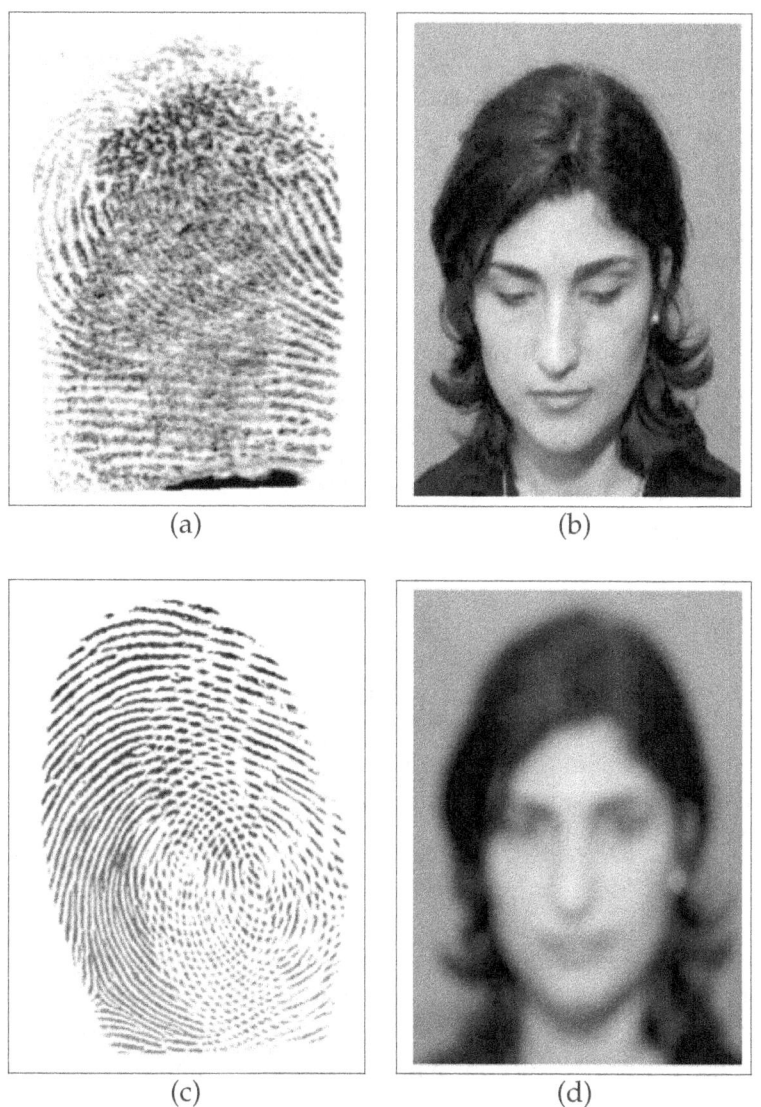

Figure 1.1: Examples of (a) a low-character fingerprint due to poor skin condition, (b) a poor-quality facial image due to poor user behavior (non-frontal pose), (c) a poor-quality fingerprint due to unclean platen with residual fingerprint on it, and (d) a low-quality (out of focus) facial image because of imperfect acquisition setting.

be stored and could be used by some predictor. The fact that quality has historically been conceived of as scalar is a widely manifested restriction. For example, BioAPI [16, 17] has a signed single byte value, BioAPI_QUALITY; and the headers of the ISO/IEC biometric data interchange format standards [18] have one byte for storing quality score. However, vector quality quantities could be used to specifically direct reacquisition attempts (e.g., camera settings) or direct enhancement of image (e.g., contrast adjustment). This is similar to the subject of "actionable" quality discussed above, with the exception that quality components are not restricted to character, behavior, imaging and environment.

Vector of quality components have been considered (e.g., the defect fields of [3]), and their practical use would require application of a discriminant function.

Quality scores are most often computed from a set of measurements made on that sample. The measurements, which can form the elements of a vector, are typically quantitative statements of how good specific properties of the image or signal data are. So, for example, the overall quality and utility of a facial image might be derived from measurements of geometric properties such as size of the face and pose angle, photometric properties such as optical resolution, compression, and saturation, and behavioral aspects such as facial expression and facial hair.

In the general case, the set of quality-related features that are germane to sample quality assessment are specific to the biometric modality and in some cases to the intended biometric matching algorithm. Thus, as discussed elsewhere, fingerprint quality may well be a function of the ability to distinguish true minutiae from spurious ones, and thus a quality vector might be comprised of elements such as local noise, compression ratio, continuity of ridge flow, area of the imaged impression, and number of minutiae. Similarly for iris recognition, quality related measurements might summarize area of the exposed iris, focus, gaze direction, and motion blur.

For any given modality, there will be a common set of specific quality problems that will degrade the accuracy of all of the biometric recognition algorithms. For example, if an image is massively compressed, all face recognition algorithms will fail. Similarly if an eyelid occludes 90% of an iris, the matching accuracy will be poor. However, the detailed behavior of matching algorithms will depend on specific factors. Thus one iris recognition algorithm may be sensitive to low contrast while another may not.

A set of measurements that constitute a quality vector will clearly convey more information than just a summary scalar value. However, two outstanding issues remain. First is that the vector in itself is not immediately useful: some operation (e.g. a mapping of the vector to a scalar) must be performed before the value can be acted upon (e.g. compared against a required minimum threshold). Thus, it may be necessary to establish a a mapping $f : \mathcal{R}^N \to \mathcal{R}^\infty$ of an N-element quality feature vector to an actionable scalar quality value. Establishing such a mapping might require a calibration procedure and an appropriate set of samples. The second issue is that the quality vectors are less interoperable than scalar values because the specific elements of the vectors are not standardized. This arises, in some part, because some biometric recognition algorithms are more sensitive to specific quality-related defects than others. For example, if a fingerprint image is rotated, this will present a serious problem to some matching algorithms but no problem to others. That means that the mapping f will need to be tailored to the recipient.

1.3 Number of Quality Values

A quality metric is more useful if operationally it may be thresholded at one of many distinct operating points. Thus a discrete-valued quality measure is better if performance is significantly different for different levels of quality. If they are not, they could be mapped to fewer levels that are statistically distinct. Real values can be quantized.

Biometric standards quite reasonably recommend quality values on the range of $[0, 100]$ with the implication that there are that many distinct values (i.e., between 6 and 7 bits). BioAPI [16, 17], for instance, specify four ranges ($[0, 25]$, $[26, 50]$, $[51, 75]$, $[76, 100]$) with associated meanings: *unacceptable, marginal, adequate* and *excellent*. This is a tacit acknowledgment that the range $[0, 100]$ is too fine, and that an integer quality value on the range $[1, 4]$ is effectively all that may be needed (or possible). Practically this may not be the case and a coarser quantization, corresponding to $L < 100$ statistically separate levels, is usually achieved. Such mapping is most accurate if provided by the author of the quality algorithm. Clearly a mathematical rationale for L (for example, a criterion against which L can be optimized) is preferable. This could be something like the knees of the distribution functions of the genuine and impostor scores, or L levels based on the separation of the two distributions. An alternative might be to let L be a free parameter in a fitting process, analogous to some discovered intrinsic precision. Regardless of how L is determined, for a quality algorithm to be effective and operationally meaningful, its L quality levels shall be statistically separate.

Chapter 2

Uses of Quality Assessments

Quality measurement plays vital role in improving biometric system accuracy and efficiency during the capture process (as a control-loop variable to initiate reacquisition), in database maintenance (sample update), in enterprise-wide quality assurance surveying, and in invocation of quality-directed processing of samples. Neglecting quality measurement will adversely impact accuracy and efficiency of biometric recognition systems (e.g., verification and identification of individuals).

Deploying quality measurement tools allows automatic quality control over biometric samples at the time of capture. If the first sample captured is of insufficient quality, it is possible to catch this in real time and request that the subject's fingerprint be retaken on the spot. Measuring quality also introduces the ability for biometric matching systems to devote different levels of computing resources according to the assessed quality of the fingerprint image. Those samples that are determined to be of low quality may be routed to slower, more robust matching algorithms, while the higher volume of high-quality samples may be routed to faster matching algorithms. Also, the weights for multimodal biometric fusion can be selected to allow better quality biometric samples to dominate the fusion. These valuable uses of BQAMs prompts the recommendation that quality values should be computed across all retained samples in an enterprise. This may be done online or offline and will depend on factors such as:

▷ the computational cost of BQAM execution during enrollment or verification;
▷ whether or not the samples are retained (in verification, they may not be);
▷ whether the matching scores or decisions themselves constitute a reportable operational performance measure; and
▷ the timescale for production of quality summaries.

Once quality scores have been collected in a central location, summarization of those scores would allow quality monitoring across multiple sites or over time. This is useful to signal possible performance problems ahead of some subsequent matching operation. Quality summarization functions should weight the native quality values so that the summarized quality value is an estimate of the expected error rate, which, for verification, should serve as measures of the overall expected false non-match rate. Arithmetic mean is not the preferred method of summarizing quality scores because all samples, regardless of their quality values, are given the same weight, but the recognition error rates are usually nonlinearly dependent on the quality values.

The following sections describe the roles of a sample quality score in the various contexts of biometric operations. The quality value here is simply a scalar summary of a sample that is taken to be some indicator of matchability.

2.1 Quality at the Time of Capture

2.1.1 Enrollment Phase Quality Assessment

Enrollment is usually a supervised process, and it is common to improve the quality of the final stored sample by acquiring as many samples as are needed to satisfy either an automatic quality measurement algorithm (the subject of this paper), a human inspector (a kind of quality algorithm), or a matching criterion (by comparison with a second sample acquired during the same session). Our focus on automated systems' needs is warranted regardless of analyses of these other methods, but we do contend that naive human judgment will only be as predictive of a matcher's performance as the human visual system is similar to the matching system's internals, and it is not evident that human and computer matching are functionally comparable. Specifically, human inspectors may underestimate performance on overtly marginal samples. Certainly human inspectors' judgment may be improved if adequate training on the failure modes and sensitivities of the matcher is given to the inspector, but this is often prohibitively expensive or time-consuming and not scalable. Immediate matching also might not be predictive of performance over time because same-session samples usually produce unrealistically high match scores. For instance, Figure 2.1 shows an example of two same-session fingerprint images that were matched successfully by three commercial vendors despite their obvious poor quality. That said, this document does not take a position on the merits of doing this.

In any case, by viewing sample acquisition as a measurement and control problem in which the control loop is closed on the quality measure, a system gains a powerful means of improving overall sample quality and therefore improving overall performance. We demonstrate this in more detail in Section 5.1, and here only report that removing lowest-quality fingerprints from an operational dataset (1.65% of the dataset) improved Equal Error Rate of a commercial fingerprint matcher from 0.0047 to 0.0024.

It is important to note that recapture will not alway improve quality. For a small fraction of the population, satisfying quality requirements (i.e., providing samples with quality

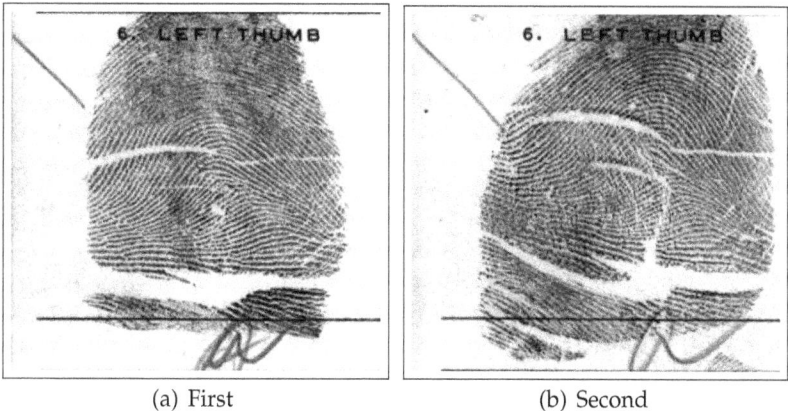

(a) First (b) Second

Figure 2.1: Example of same-session captures of single finger that, despite their poor quality (NFIQ =5), were matched correctly by three leading commercial matchers.

better than some pre-defined threshold) might not be achievable, because the source of biometrics (e.g., surface of finger skin) is impaired, for example, due to age. These are subjects of *low biometric character*, and recapture will mostly not improve the quality (see Section 1.1 for a discussion of quality aspects). Not satisfying quality requirements results in failure to enroll the subject. Failure to enroll cases need additional processing (e.g., different capture device), which is costly. Therefore, when determining quality threshold at the time of capture, it is important to take failure to enroll rate into consideration.

In Appendix ??, we recommend a procedure to set quality threshold at time of capture for both enrollment and verification or identification phases.

2.1.2 Verification Phase Quality Assessment

During a verification transaction, quality can be improved by closing an acquire-reacquire loop on either a match-score from comparison of new and enrollment samples or on a quality value generated without matching. Indeed, it is common to implement an "up to three attempts" policy in which a positive match is a *de facto* statement that the sample was of good quality - even if the individual happens to be an impostor. Depending on the relative computational expenses of sample matching, reacquisition, and quality measurement, the immediate use of a matcher may not be the best solution.

The key difference here (as compared to the enrollment phase) is that quality values of both the enrollment and verification samples can be used to predict performance. This two-dimensional problem is distinct from the enrollment case where only one quality value is used. Just like the enrollment quality assessment, when setting quality threshold for the verification phase, failure to acquire should be considered (see Appendix ??).

2.1.3 Identification Quality Assessment

Quality measurement in identification systems is important for at least three reasons. First, many users often do not have an associated enrollment sample. So a one-to-many match will be an inefficient and inconclusive method of stating whether the authentication sample had high quality. Second, in negative identification systems where users with an enrolled sample are motivated to evade detection, quality measurement can be used to detect and prevent submission of samples likely to perform poorly [13], which may help prevent attempts at spoofing or defeating detection. Third, identification is a difficult task; it is imperative to minimize both the false non-match rate (FNMR) *and* the false match rate (FMR). To the extent that consistently high-quality samples will produce high genuine scores, a high matching threshold can be used, and this will collaterally reduce FMR . But in large populations, FMR becomes dominant, and this raises the question: can a quality apparatus be trained to be directly predictive of false match likelihood? At the minimum, the ability to maintain a fairly constant FMR regardless of the quality of images is a desirable feature. Increase in FMR when quality degrades is a security vulnerability. It increases the chance of an impostor spoofing the system by presenting poor-quality samples.

2.2 Performance-Related Quality Monitoring

A valuable use of BQAMs is to monitor quality across multiple sites or over time. Quality values may be aggregated and compared with some historical or geographic baselines. Use of quality values in this role has been documented in [1]. Therefore, there is a need for procedures to summarize quality values computed across all retained samples in an enterprise into a single quality value representing the overall quality of the enterprise. Quality summarization supports monitoring

- over time (to expose seasonal variation or trends);
- for each sensor (to identify defective devices);
- at each site (to identify problem locations);
- of officials or attendants (to assess adherence to operating procedures); and
- per user basis (to identify users that consistently yield low-quality samples).

In each case, the quality summaries can be used to identify departures from the application-specific historical norms, or design targets. Once quality values have been collected in a central location, these should be aggregated. The result is a summary value which supports monitoring of quality. Quality summarization should be performed across similar usage, e.g., quality summarization over all enrollment samples of an enterprise, or quality summarization over all verification samples of an enterprise. In operations where users frequently interact with a biometric system (e.g., time and attendance applications), quality values may be aggregated on a per-user basis. This will reveal the existence of indviduals that consistently yield low-quality samples.

In Section 7.2, we show that it is generally not sufficient to simply average those values. Instead we recommend that the provider of a quality assessment algorithm should supply a function to aggregate values into a summary statistic. For verification applications, quality summarization functions should weight the native quality values to reflect mean expected false non-match rate (FNMR). Furthermore, it is recommended that such functions compute quality summaries on the standardized range of biometric sample quality values as specified in ISO/IEC 19784-1 BioAPI [17], which requires single-sample quality values on $[0, 100]$.

The quality summarization function could be the result of a BQAM calibration process conducted by the provider, by a third party laboratory, or by the deploying organization.*in-situ*[1]

The recommended procedure for National Institute of Standards and Technology (NIST) Fingerprint Image Quality NFIQ [6,7] is given in Section 7.2. This kind of quality aggregation applied here to NFIQ may be appropriate for other quality measures. However, this document does not prescribe any particular functional form, and developers are free to use any appropriate method. Indeed, we anticipate (and encourage) that such methods will remain the private intellectual property of the provider.

Table 2.1 shows example of quality (NFIQ) summarization across different sites and per users. NFIQ summarization was performed on fingerprints collected as part of the US-VISIT program over one week.

2.3 Differential Processing

Quality measurement algorithms can be used to alter the subsequent processing of a sample. Such conditional activity are categorized as follows.

> **Pre-processing Phase** A biometric recognition system might apply image restoration algorithms (e.g., contrast adjustment) or invoke different feature extraction algorithms for samples with some discernible quality problem.
> **Matching Phase** Certain systems may invoke a slower but more powerful matching algorithm when low-quality samples are compared. Note that use of the slower, more accurate matcher for *all* the samples would greatly increase the processing time.

---

[1]A representative set of (mated) samples and one or more matching algorithms will be needed for calibration.

Table 2.1: NFIQ summary of a random selection of fingerprint images collect as part of the US-VISIT program at two different locations over one week.

# of samples	consulate-1	consulate-2
total # of samples	1431	4768
# of samples of NFIQ 1	939	831
# of samples of NFIQ 2	247	197
# of samples of NFIQ 3	84	184
# of samples of NFIQ 4	12	39
# of samples of NFIQ 5	3	15
NFIQ summary	98.13	95.50
confidence interval	(97.68,9 8.47)	(94.95, 96.35)

Routing only poor-quality samples to the slower but more accurate matching algorithm improves the overall accuracy without negatively impacting the efficiency of the system.

▷ **Decision Phase** The logic that renders acceptance or rejection decisions may depend on the measured quality of the original samples. This might involve changing a verification system's operating threshold for poor-quality samples. In multimodal biometrics, the relative qualities of samples of the separate modes may be used to augment a fusion process [14, 15].

▷ **Sample Replacement** To negate the effects of template aging, a quality measurement may be used to determine whether a newly acquired sample should replace the enrolled one. An alternative would be to retain both the old and new samples for use in a multi-instance fusion scheme.

▷ **Template Update** Again, to address template aging, some systems instead combine old and new sample features. Quality could be used in this process.

Chapter 3

Standardization

This chapter focuses on biometric quality standardization. Broadly, biometric quality standards serve the same purpose as many other standards, which is to establish an interoperable definition, interpretation, and exchange of biometric quality data. Like other standards, this creates grounds for a marketplace of off-the-shelf products, and is a necessary condition to achieve supplier independence and to avoid vendor lock-in.

With advancement in biometric technologies as a reliable identity authentication scheme, more large-scale deployments (e.g., e-passport) involving multiple organizations and suppliers are being rolled out. Therefore, in response to a need for interoperability, biometric standards have been developed.

Without interoperable biometric data standards, exchange of biometric data among different applications is not possible. Seamless data sharing is essential to identity management applications when enrollment, capture, searching, and screening are done by different agencies, at different times, using different equipment in different environments and/or locations. Interoperability allows modular integration of products without compromising architectural scope, and facilitates the upgrade process and thereby mitigates against obsolescence.

Biometric data interchange standards are needed to allow the recipient of a data record to successfully process data from an arbitrary producer. This defines biometric interoperability and the connotation of the phrase "successfully process" is that the data, in this case, biometric quality score, can be accurately exchanged and interpreted by different applications. This can be achieved only if the data record is both syntactically and semantically conformant to the documentary standard.

Standards do not in and of themselves assure interoperability. Specifically, when a standard is not fully prescriptive or it allows for optional content, then two implementations that are both exactly conformant to the standard may still not interoperate. This situation may be averted by applying further constraints on the application of the standard. This is done by means of "application profile" standards, which formally call out the needed base standards and refine their optional content and interpretation.

3.1 The ISO/IEC 29794 Biometric Sample Quality Standard

In January 2006, the Biometrics Subcommittee (SC 37) of Joint Technical Committee (JTC 1) initiated work on ISO/IEC 29794, a multipart standard that establishes quality requirements for generic aspects (Part 1), fingerprint image (Part 4), facial image (Part 5), and, possibly, other biometrics later. Specifically, part 1 of this multi part standard specifies derivation, expression, and interpretation of biometric quality regardless of modality. It also addresses the interchange of biometric quality data via the multipart ISO/IEC 19794 Biometric Data Interchange Format Standard. Parts 4 and 5 are technical reports (not standard drafts) which address aspects of biometric sample quality that are specific to finger images and facial images as defined in ISO/IEC 19794-4 and ISO/IEC 19794-5, respectively.

The generic ISO quality draft (ISO/IEC 29794-1) requires that quality values must be indicative of recognition performance and considers three components of biometric sample quality, namely, character, fidelity, and utility, as shown in Figure 3.1. The character of a sample indicates the richness of features and traits from which the biometric sample is derived. The fidelity of a sample is defined as the degree of similarity between a biometric sample and its source; for example, a heavily compressed fingerprint has low fidelity. The utility of a sample reflects the observed or predicted positive or negative contribution of an individual sample to the overall performance of a biometric system. Utility is a function of both the character and fidelity of a sample and is most closely indicative of performance in terms of recognition error rates.

Part 1 of multipart ISO/IEC 29794 draft standard defines a binary record structure for the storage of a sample's quality data. It establishes requirements on the syntax and semantic content of the structure. Specifically, it states that the purpose of assigning a quality score to a biometric sample shall be to indicate the expected utility of that sample in an automated comparison environment. That is, a quality algorithm should produce quality scores that target application-specific performance variables. For verification, the metric would usually be false-match and false-non-match rates that are likely to be realized when the sample is matched.

In addition, revision of all parts of ISO/IEC 19794 Biometric Data Interchange Format started in January 2007. This opened the opportunity to revise or add quality-related clauses (e.g., compression limits) to data format standards so that conformance to those standards ensures acquisition of sufficient quality samples. This constitutes quality by design. To enable an interoperable way of reporting and exchanging biometric data quality scores, the inclusion of a five-byte quality field to the view header in each view of the data in a Biometric Data Block (BDB) for all parts of ISO/IEC 19794 is being considered. By placing quality field in the view header (as opposed to general header) of a BDB, one can precisely report the quality score for each view of a biometric sample (see Figure 3.2). Table 3.1 shows the structure of the quality filed that SC 37 Working Group 3 is currently considering.

The one-byte quality score shall be a quantitative expression of the predicted matching performance of the biometric sample. Valid values for quality score are integers between 0 and 100, where higher values indicate better quality. Values of 254 and 255 are to handle special cases. An entry of "255" shall indicate a failed attempt to calculate a quality score, and an entry of "254" shall indicate no attempt was made to calculate a quality score (i.e., no quality score has been specified). These values of quality score are harmonized with ISO/IEC 19784-1 BioAPI Specification (Section 3.3) [17], where "255" is equivalent to BioAPI "-1" and "254" is equivalent BioAPI "-2" (Note that BioAPI, unlike ISO/IEC 19794, uses signed integers.).

To enable the recipient of the quality score to differentiate between quality scores generated by different algorithms, the provider of quality scores shall be uniquely identified by the two most significant bytes of four-byte Quality Algorithm vendor ID (QAID). The least significant two bytes shall specify an integer product code assigned by the vendor of the quality algorithm. It indicates which of the vendors algorithms (and version) was used in the calculation of the quality score and should be within the range 1 to 65535. Quality Algorithm Vendor ID shall be set to "0" if the Quality Score is 254.

The structure of the quality field is modality-independent and therefore generalizable to all parts of ISO/IEC 19794.

The ISO/IEC 29794 standard is currently under development, and ISO/IEC 19794 is currently under revision. The reader is cautioned that standards under development or revision are subject to change; the documents are owned by the respective working groups and their content can shift due to various reasons including, but not limited to, technical difficulties, the level of support, or the need to gain consensus.

3.2 The ANSI/NIST ITL 1-2007 Quality Field

Initiated in 1986, this standard is the earliest and most widely deployed biometric standard. It establishes formats for the markup and transmission of textual, minutia, and image data between law enforcement agencies, both within the United States and internationally.

The ANSI/NIST standard includes defined *Types* for the major biometric modalities. The standard is multimodal in that it allows a user to define a transaction that would require, for example, fingerprint data as Type 14, a facial mugshot as Type 10, and the mandatory

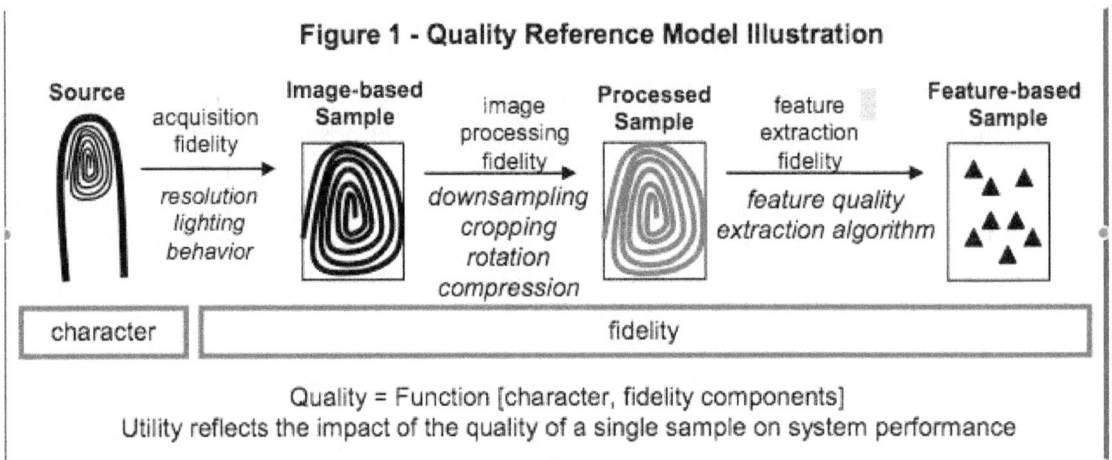

Figure 3.1: Components of quality as defined by ISO/IEC 29794 Biometric Sample Quality - Part 1: Framework. The character of a sample indicates the richness of features from which the biometric sample is derived. The fidelity of a sample is the degree of similarity between a biometric sample and its source. The utility of a sample is indicative of positive or negative contribution of an individual sample to the overall performance of a biometric system. Source: ISO/IEC JTC1 SC 37 N2727.

Figure 3.2: Structure of header in a biometric data block as defined in ISO/IEC 19794-x.

Table 3.1: Structure of five-byte quality field that SC 37 Working Group 3 is considering.

DESCRIPTION	SIZE	VALID VALUES	NOTE
Quality Score	1 byte	[0-100] 254, 255	0: lowest
Quality Algorithm Vendor ID	2 bytes	0 if Quality Score = 254 [1,65535] otherwise	These two bytes uniquely identify the supplier (vendor) of quality score.
Quality Algorithm ID	2 bytes	0 if Quality Score = 254 [1,65535] otherwise	These two bytes uniquely identify the algorithm that computes the quality score. It is provided by the supplier (vendor) of quality score.

header and metadata records Type 1 and 2. These are linked with a common numeric identifier.

In its latest revision [28], the standard adopted the ISO five-byte quality field (Table 3.1) structure, but unlike ISO/IEC 29794, it allows for multiple quality fields, where each quality score could be computed by a different quality algorithm supplier. In addition, it mandates NIST Fingerprint Image Quality (NFIQ) [7] for all Type 14 records.

3.3 The BioAPI Quality Specification

ISO/IEC 19784 Biometric Application Programming Interface (BioAPI) [17] (and its national counterpart The BioAPI specification [16]) allows for quality measurements as an integral value in the range 0-100, with exceptions that value of "-1" means that quality field was not set by the Biometric Service Provider (BSP) and value of "-2" means that quality information is not supported by the BSP. The primary objective of quality measurement and reporting is to have the BSP inform the application how suitable the biometric sample is for the purpose specified by the application (as intended by the BSP implementer based on the use scenario envisioned by that BSP implementer), and the secondary objective is to provide the application with relative results (e.g., current sample is better/worse than previous sample). BioAPI also provides guidance on general interpretation of quality scores as shown in Table 3.2).

Value	Interpretation
0-25	UNACCEPTABLE: The sample cannot be used for the purpose specified by the application. The sample needs to be replaced using one or more new biometric samples.
26-50	MARGINAL: The sample will provide poor performance for the purpose specified by the application and, in most application environments, will compromise the intent of the application. The sample needs to be replaced using one or more new biometric samples.
51-75	ADEQUATE: The biometric data will provide good performance in most application environments based on the purpose specified by the application. The application should attempt to obtain higher-quality data if the application developer anticipates demanding usage.
76-100	EXCELLENT: The biometric data will provide good performance for the purpose specified by the application.

Table 3.2: BioAPI quality categories

Chapter 4

Properties of a Quality Measure

This chapter gives needed background material, including terms, definitions, and data elements, for later chapters on quality evaluation, annotation, and summarization.

Throughout this chapter, we use low-quality values to indicate poor sample properties as suggested by Biometric Quality Standards [17, 18]. This is at odds with some systems (for example, the NIST Fingerprint Image Quality (NFIQ) algorithm [6]), for which low values indicate good "quality". Accordingly, this document transforms the raw NFIQ values $1 \ldots 5$ using $Q = 6-\text{NFIQ}$.

4.1 Notation

Consider a data set D containing two samples, $d_i^{(1)}$ and $d_i^{(2)}$ collected from each of $i = 1, \ldots, N$ individuals. The first sample can be regarded as an enrollment image, the second as a user sample collected later for verification or identification purposes. The appropriate composition of this data set for quality measurement algorithm assessment is discussed later in Section 6. Suppose a quality algorithm Q can be run on the i-th enrollment sample to produce a quality value

$$q_i^{(1)} = Q(d_i^{(1)}) \qquad (4.1)$$

and likewise for the authentication (use-phase) sample

$$q_i^{(2)} = Q(d_i^{(2)}) \qquad (4.2)$$

We formalize our premise that biometric quality measures should predict performance. That is, we formalize quality values q_i are related to recognition error rates. A formal statement of such requires an appropriate, relevant, and tractable definition of performance, which is given below.

4.2 Relationship to Matching

Consider K verification algorithms, V_k, that compare pairs of samples (or templates derived from them) to produce match (i.e., genuine) similarity scores

$$s_{ii}^{(k)} = V_k(d_i^{(1)}, d_i^{(2)}) \qquad (4.3)$$

and similarly non-match (impostor) scores

$$s_{ij}^{(k)} = V_k(d_i^{(1)}, d_j^{(2)}) \quad i \neq j. \qquad (4.4)$$

If we now posit that two quality values can be used to produce an estimate of the genuine similarity score that matcher k would produce on two samples

$$s_{ii}^{(k)} = P(q_i^{(1)}, q_i^{(2)}) + \epsilon_{ii}^{(k)} \qquad (4.5)$$

where the function P is some predictor of a matcher k's similarity scores, and ϵ_{ii} is the error in doing so for the i-th score. Substituting equation (4.1) gives

$$s_{ii}^{(k)} = P(Q(d_i^{(1)}), Q(d_i^{(2)})) + \epsilon_{ii}^{(k)} \qquad (4.6)$$

and it becomes clear that together P and Q would be perfect imitators of the matcher V_k in equation (4.3) if it was not necessary to apply Q to the samples separately. This separation is usually a necessary condition for a quality algorithm to be useful because at least half of the time (i.e., enrollment) only one sample is available (see Section 2). Thus the quality problem is hard; first, because Q is considered to produce a scalar, and secondly, because it is applied separately to the samples. The obvious consequence of this formulation is that it is inevitable that quality values will imprecisely map to similarity scores, i.e., there will be scatter of the known scores, s_{ii}, for the known qualities $q_i^{(1)}$ and $q_i^{(2)}$. For example, Figure 4.1 shows the raw similarity scores from a commercial fingerprint matcher versus the transformed integer quality scores from the NFIQ algorithm [5], where NFIQ native scores are mapped to $Q = 6-$NFIQ. Figure 4.1(a) also includes a least squares linear fit, and Figure 4.1(b) shows a cubic spline fit of the same data. Both trend in the correct direction: worse quality gives lower similarity scores. However, even though the residuals in the spline fit are smaller than those for the linear, they still are not small. Indeed, even with a function of arbitrarily high order, it will not be possible to fit the observed scores perfectly if quality values are discrete (as they are for NFIQ). By including the two fits of the raw data, we do not assert that scores should be linearly related to the two quality values (and certainly not locally cubic). Accordingly, we conclude that it is unrealistic to require quality measures to be linear predictors of the similarity scores; instead, the scores should be a monotonic function, that is, higher-quality samples give higher-similarity scores.

4.3 Relationship to Performance

Quality measurement algorithms should be designed to target application-specific performance variables. For verification, these would be the false match or false non-match rates. For identification, the metrics would usually be FNMR and FMR [19], but these may be augmented with rank and candidate-list length criteria. Closed-set identification is operationally rare and is not considered here.

Verification is a positive application, which means samples are captured overtly from users who are motivated to submit high-quality samples. For this scenario, the relevant performance metric is the false non-match rate (FNMR) for genuine users, because two high-quality samples from the same individual should produce a high score. For FMR, it should be remembered that false matches should occur only when samples are biometrically similar (with regard to a matcher) as, for example, when identical twins' faces are matched. So high-quality images should give very low impostor scores, but low-quality images should also produce low scores. Indeed, it is an undesirable trait for a matching algorithm to produce high impostor scores from low-quality samples. In such situations, quality measurement should be used to preempt submission of a deliberately poor sample (see the uses discussion in Section 2).

For identification, FNMR is of primary interest. It is the fraction of enrollee searches that do not yield the matching entry on the candidate list. At a fixed threshold, FNMR is usually considered independent of the size of the enrolled population, because it is simply dependent on one-to-one genuine scores. However, because impostor acceptance, as quantified by FMR, is a major problem in identification systems, it is necessary to ascertain whether low- or high-quality samples tend to cause false matches.

For a quality algorithm to be effective, an increase in FNMR and FMR is expected as quality degrades. The plots in Figure 4.2 show the relationship of transformed NFIQ quality levels to FNMR and FMR. Figure 4.2(a) and 4.2(c) are boxplots of the raw genuine and impostor scores for each of the five quality levels. The scores were obtained by applying a commercial fingerprint matcher to left and right index finger impressions of 34,800 subjects. Also shown are boxplots of FNMR and FMR. The result, that the two error rates decrease as quality improves, is expected and beneficial. The FMR shows a much smaller decline. The non-overlap of the notches in plots of 4.2(a) and 4.2(b) demonstrates "strong evidence" that the medians of the quality levels differ [27]. If the BQAM had more finely quantized its output, to $L > 5$ levels, this separation would eventually disappear (see discussion in Section 1.3).

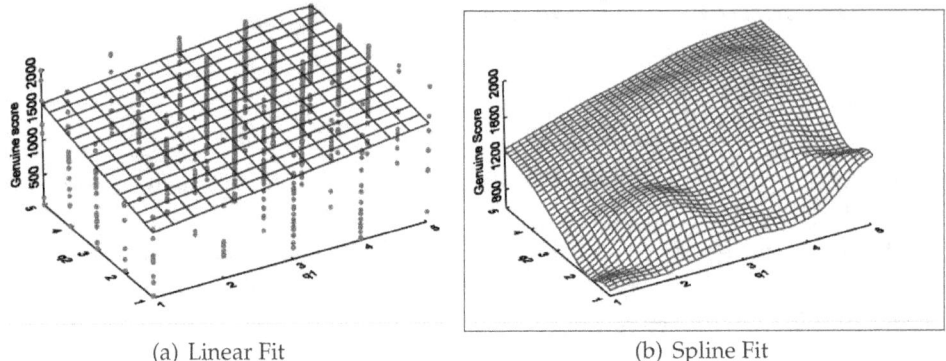

(a) Linear Fit (b) Spline Fit

Figure 4.1: Dependence of raw genuine scores on the transformed NFIQ qualities of the two input samples.

4.4 Combining Quality Values

Biometric matching involves at least two samples, and the challenge is then to relate performance to quality values $q^{(1)}$ and $q^{(2)}$. This empirical dependence of performance on two values was shown in Figure 4.1. We simplify the analysis by combining the two qualities

$$q_i = H(q_i^{(1)}, q_i^{(2)}) \tag{4.7}$$

As discussed in Section 2, it is usually the case that operationally a BQAM can be used to ensure that an enrollment sample is of high quality. This will be compared later with a sample that typically is of less controlled quality. To capture this concept, we consider $H(x, y) = \min(x, y)$, i.e., the worse of two samples drives the similarity score. Some other relevant pair-wise combination function H includes (but certainly is not limited to) the arithmetic and geometric means, $H(x, y) = (x + y)/2$ and $H(x, y) = \sqrt{xy}$ (see [20]), and the difference function $H(x, y) = |x - y|$. We note that whatever H is used, it should be well-defined for allowed values of x and y (e.g., positive values for the geometric mean).

Figure 4.3 shows error vs. reject behavior for the NFIQ quality method when the various $H(q_1, q_2)$ combination functions are used. Error vs. reject shows the improvement in performance (FNMR or FMR) as samples of lowest quality are rejected. Error vs. reject concept is discussed in detail in Section 5.2.

The lines in Figure 4.3 show $H(q_1, q_2) + N(0, 0.01)$, where the gaussian noise serves to randomly reject samples within a quality level and produces an approximation of the lower convex hull of the geometric mean curve. Between the minimum, mean, and geometric mean functions, there is little difference. The yellow line result, for $H = |q_1 - q_2|$, shows that transformed genuine comparison score is unrelated to the difference in the qualities of the samples. Instead, the conclusion is that FNMR is related to monotonic functions of the two values. The applicability of this result to other quality methods is not known.

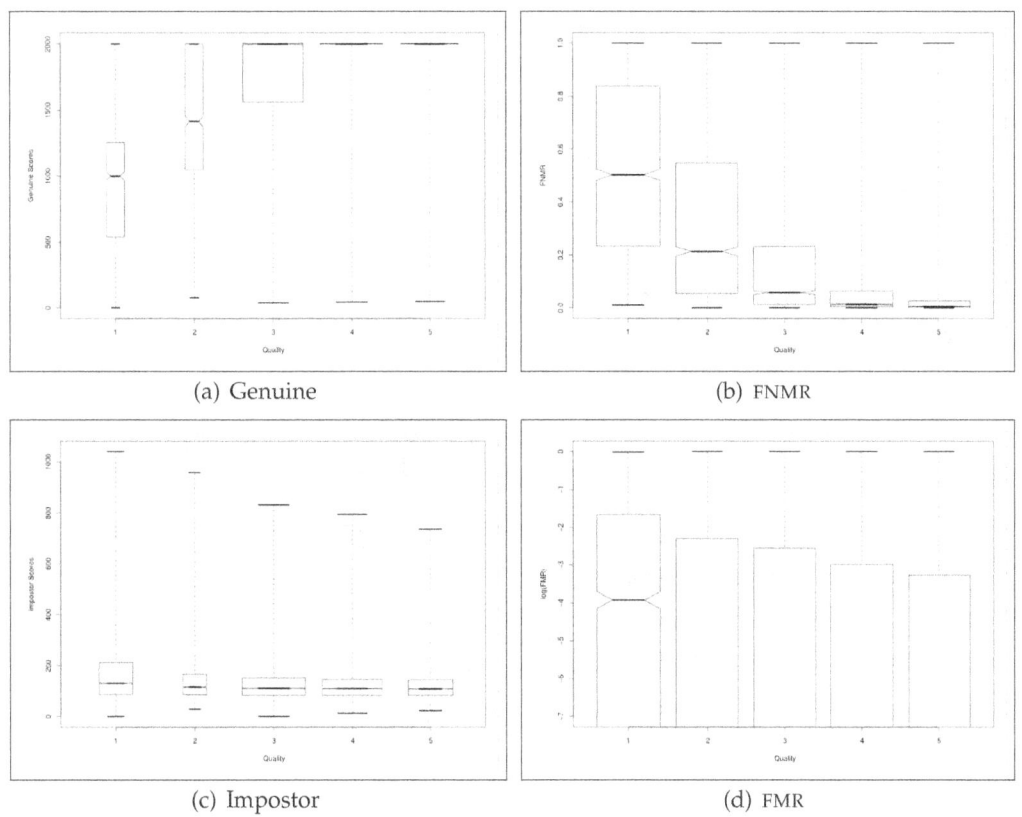

Figure 4.2: Boxplots of genuine scores, FNMR, impostor scores, and FMR for each of five transformed NFIQ quality levels for scores from a commercial matcher. Each quality bin, q, contains scores from comparisons of enrollment images with quality $q^{(1)} \geq q$ and subsequent use-phase images with $q^{(2)} = q$, per the discussion in Section 5.1. The boxplot notch shows the median, the box shows the interquartile range, and the whiskers show the extreme values. Notches in (d) are not visible, because the medians of FMR s are zero and therefore outside the plot range.

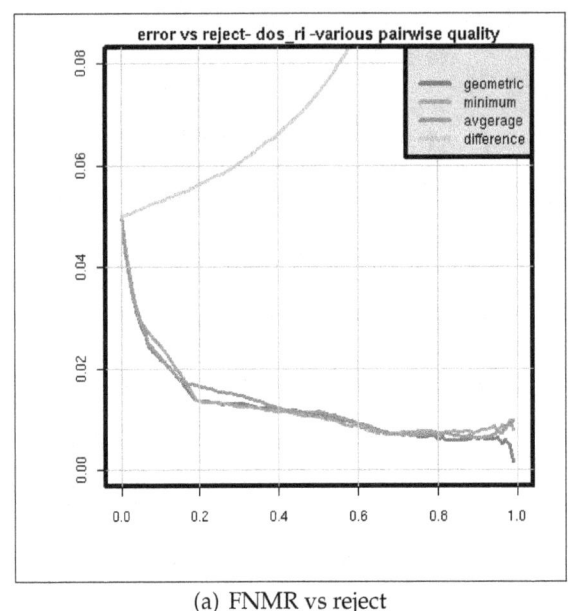

(a) FNMR vs reject

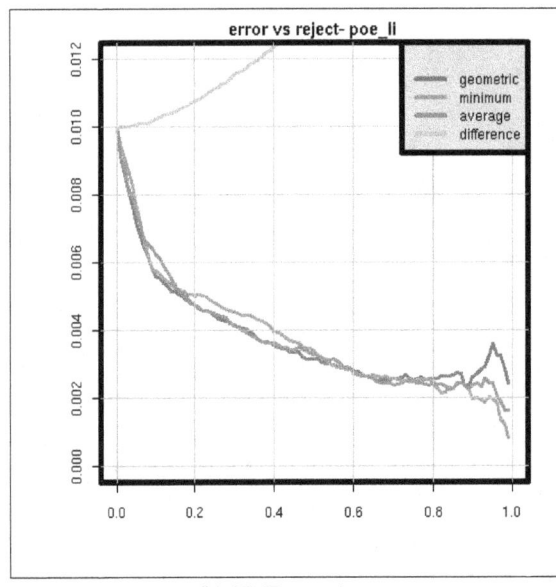

(b) FMR vs reject

Figure 4.3: Dependence of the error vs. reject characteristic on the quality combination function H(.). (a) shows, for a fixed threshold, the decrease in FNMR as samples with low-quality scores are rejected. (b) shows the decrease in FMR at a fixed threshold, as samples with low-quality scores are rejected. Index fingerprints collected at operational settings were used. The similarity scores come from commercial matchers.

Chapter 5

Do Quality Values Predict Performance?

This chapter documents methods for the quantitative evaluation of systems that produce a scalar summary of a biometric sample's quality. Quality measurement algorithm is regarded as a black box that converts an input sample to an output scalar. Evaluation should be done by quantifying the association between those values and observed matching results. We advance detection error trade-off and error versus reject characteristics as metrics for the comparative evaluation of sample quality measurement algorithms.

Prior work on quality evaluation, and of sample quality analysis generally, is limited. Quality measurement naturally lags recognition algorithm development, but has emerged as it is realized that biometric systems fail on certain pathological samples. The primary use of a quality measure is as a means of detecting a bad sample and initiating recapture of the live subject. "Bad" in this context, refers to any property or defect associated with a sample that would cause performance degradation.

We recommend testing quality measurement algorithms in large-scale offline trials which offer repeatable, statistically robust means of evaluating core algorithmic capability. Alonso et al. [9] reviewed five algorithms and used the fingerprints of the multimodal MCYT corpus [10] to compare the distributions of the algorithms' quality assignments, with the result that most of the algorithms behave similarly. We note that finer-grained aspects of sample quality can be addressed. For instance, Lim et al. [11] trained a fingerprint quality system to predict the accuracy of minutia detection. However, such methods rely on the manual annotation of a data set, and this is impractical for all but small datasets, not least because human examiners will disagree in this respect. The virtue of relating quality to performance is that matching trials can be automated and conducted in bulk. We note further that quality algorithms that relate to human perception of a sample quantify performance only as much as the sensitivities of the human visual system are the same as those of a biometric matcher. One further point is that performance-related quality evaluation is agnostic on the underlying technology: it would be improper to force a fingerprint quality algorithm to produce low-quality values for an image with few minutia when the target matching algorithm is non-minutia-based, as is the case for pattern-based methods [12].

The evaluation protocols proposed assume only that the quality algorithm is claimed to predict performance: we do not assume that the algorithm has been standardized nor that its output has any particular distribution. We test the claim by relating quality values to empirical matching results. However, we consider the algorithm to be a black box whose design and intended outputs are determined solely by its author, and we make no assumption of its internal operation.

Formal specification of how performance should be quantified and whether such performance measures are viable and appropriate were discussed in section 4.3. Next sections describe three methods for the evaluation of quality. All three consider the use of combination functions, H (discussed in section 4.4), which are specifically compared in section 5.2.

5.1 Rank-Ordered Detection Error Trade-off Characteristics

A quality algorithm is useful if it can at least give an ordered indication of an eventual performance. For example, for L discrete quality levels, there should notionally be L DET characteristics.[1] In the studies that have evaluated quality measures [4, 6, 19, 20, 25, 26], DET's are the primary metric. We recognize that DETs are widely understood, even expected, but note three problems with their use: being parametric in threshold, t, they do not show the dependence of FNMR (or FMR) with quality at fixed t, they are used without a test of the significance of the separation of L levels; and partitioning of the data for their computation is under-reported and non-standardized.

We examine three methods for the quality-ranked DET computation. All three use N paired matching images with integer qualities $q_i^{(1)}$ and $q_i^{(2)}$ on the range $[1, L]$. Associated with these are N genuine similarity scores, s_{ii}, and up to $N(N-1)$ impostor scores, s_{ij} where $i \neq j$, obtained from some matching algorithm. All three methods compute a DET characteristic for each quality level k. For all thresholds s, the DET is a plot of FNMR (s) = $M(s)$ versus FMR $(s) = 1 - N(s)$, where the empirical cumulative distribution functions $M(s)$ and $N(s)$ are computed, respectively, from sets of genuine and impostor scores.

The three methods of partitioning differ in the contents of these two sets. The simplest case uses scores obtained by comparing authentication and enrollment samples whose qualities are both k. This procedure (see for example, [21]) is common but overly simplistic. By

---

[1]The DET used here plots FNMR vs. FMR on log scales. It is unconventional in that it does not transform the data by the CDF of the standard normal distribution. The receiver operating characteristic plots 1−FNMR on a linear scale instead. These characteristics are used ubiquitously to summarize verification performance.

plotting

$$\text{FNMR}(s, k) = \frac{\left|\left\{s_{ii}: s_{ii} \leq s, \; q_i^{(1)} = q_i^{(2)} = k\right\}\right|}{\left|\left\{s_{ii}: s_{ii} < \infty, \; q_i^{(1)} = q_i^{(2)} = k\right\}\right|} \quad (5.1)$$

$$\text{FMR}(s, k) = \frac{\left|\left\{s_{ij}: s_{ij} > s, \; q_i^{(1)} = q_j^{(2)} = k, i \neq j\right\}\right|}{\left|\left\{s_{ij}: s_{ij} > -\infty, \; q_i^{(1)} = q_j^{(2)} = k, i \neq j\right\}\right|}$$

the DETs for each quality level can be compared. Although a good BQAM will exhibit an ordered relationship between quality and error rates, this DET computation is not operationally representative because an application cannot usually accept only samples with one quality value. Rather the DET may be computed for verification of samples of quality k with enrollment samples of quality greater than or equal to k,

$$\text{FNMR}(s, k) = \frac{\left|\left\{s_{ii}: s_{ii} \leq s, \; q_i^{(1)} \geq k, q_i^{(2)} = k\right\}\right|}{\left|\left\{s_{ii}: s_{ii} \leq \infty, \; q_i^{(1)} \geq k, q_i^{(2)} = k\right\}\right|} \quad (5.2)$$

$$\text{FMR}(s, k) = \frac{\left|\left\{s_{ij}: s_{ij} > s, \; q_i^{(1)} \geq k, q_j^{(2)} = k, i \neq j\right\}\right|}{\left|\left\{s_{ij}: s_{ij} > -\infty, \; q_i^{(1)} \geq k, q_j^{(2)} = k, i \neq j\right\}\right|}$$

we model the situation in which the enrollment samples are at least as good as the authentication (i.e., user submitted) samples. Such a use of quality would lead to failures to acquire for the low-quality levels.

If instead we compare performance across *all* authentication samples against enrollment samples of quality greater than or equal to k,

$$\text{FNMR}(s, k) = \frac{\left|\left\{s_{ii}: s_{ii} \leq s, \; q_i^{(1)} \geq k\right\}\right|}{\left|\left\{s_{ii}: s_{ii} < \infty, q_i^{(1)} \geq k\right\}\right|} \quad (5.3)$$

$$\text{FMR}(s, k) = \frac{\left|\left\{s_{ij}: s_{ij} > s, \; q_i^{(1)} \geq k, i \neq j\right\}\right|}{\left|\left\{s_{ij}: s_{ij} > -\infty, \; q_i^{(1)} \geq k, i \neq j\right\}\right|}$$

we model the situation where quality control is applied only during enrollment. If repeated enrollment attempts fail to produce a sample with quality above some threshold, a failure-to-enroll (FTE) would be declared. This scenario is common and possible because enrollment, as an attended activity, tends to produce samples of better quality than authentication.

The considerable differences between these three formulations are evident in the DETs of Figure 5.1 for which the NFIQ algorithm [5] for the predicting performance of a commercial fingerprint system was applied to over 61,993 genuine and 121,997 impostor comparisons (NFIQ native scores were transformed to $Q = 6 - \text{NFIQ}$). In all cases, the ranked separation of the DETs is excellent across all operating points. We recommend that equation (5.2), as shown in Figure 5.1(b), be used because it is a more realistic operational model.

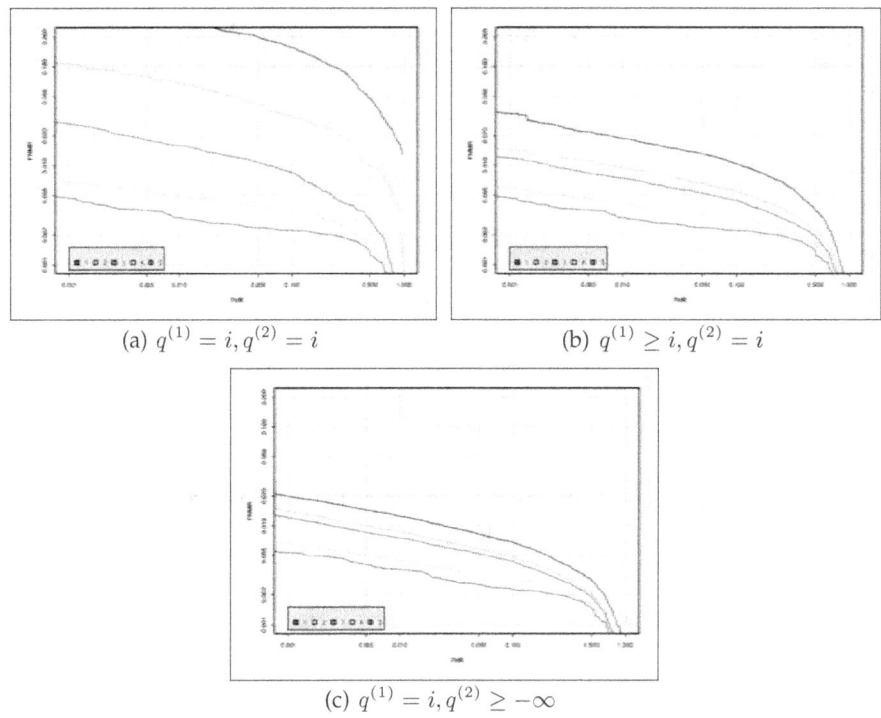

Figure 5.1: Quality-ranked detection error trade-off characteristics. Each plot shows five traces corresponding to five transformed NFIQ levels.

However, as relevant as DET curves are to expected performance, we revisit here a very important complication. Because DET characteristics quantify the separation of the genuine and impostor distributions and combine the effect of quality on both genuine and impostor performance, we lose sight of the separate effects of quality on FNMR and FMR .

In any case, we conclude that DETs, while familiar and highly relevant, confound genuine and impostor scores. The alternative is to look at the specific dependence of the error rates on quality at some fixed threshold. Indeed, for verification applications, the variation in FNMR with quality is key because the majority of transactions are genuine attempts. For negative identification systems (e.g., watchlist applications) in which users are usually not enrolled, the variation of FMR with quality is critical. This approach is followed in the next section.

5.2 Error vs. Reject Curves

In this section, we recommend to use error vs. reject curves as an alternative means of evaluating BQAMs. The goal is to state how efficiently rejection of low-quality samples results in improved performance. This again models the operational case in which quality is maintained by reacquisition after a low-quality sample is detected. Consider that a pair of samples (from the same subject), with qualities $q_i^{(1)}$ and $q_i^{(2)}$, are compared to produce a score $s_{ii}^{(k)}$, and this is repeated for N such pairs.

We introduce thresholds u and v that define levels of acceptable quality and define the set of low-quality entries as

$$R(u,v) = \left\{ j \ : \ q_j^{(1)} < u, \quad q_j^{(2)} < v \right\} \tag{5.4}$$

The FNMR is the fraction of genuine scores below threshold computed for those samples *not* in this set

$$\text{FNMR}\,(t, u, v) = \frac{|\{s_{jj} : s_{jj} \leq t, j \notin R(u,v)\}|}{|\{s_{jj} : s_{jj} < \infty, j \notin R(u,v)\}|} \tag{5.5}$$

The value of t is fixed[2] and u and v are varied to show the dependence of FNMR on quality.

For the one-dimensional case when only one quality value is used (see Section 4.4), the rejection set is

$$R(u) = \left\{ j \ : \ H(q_j^{(1)}, q_j^{(2)}) < u \right\} \tag{5.6}$$

FNMR is false non-match performance as the proportion of non-excluded scores below the threshold.

$$\text{FNMR}\,(t, u) = \frac{|\{s_{jj} : s_{jj} \leq t, j \notin R(u)\}|}{|\{s_{jj} : s_{jj} < \infty, j \notin R(u)\}|} \tag{5.7}$$

---

[2]Any threshold may be used. Practically it will be set to give some reasonable false non-match rate, f, by using the quantile function the empirical cumulative distribution function of the genuine scores, $t = M^{-1}(1 - f)$.

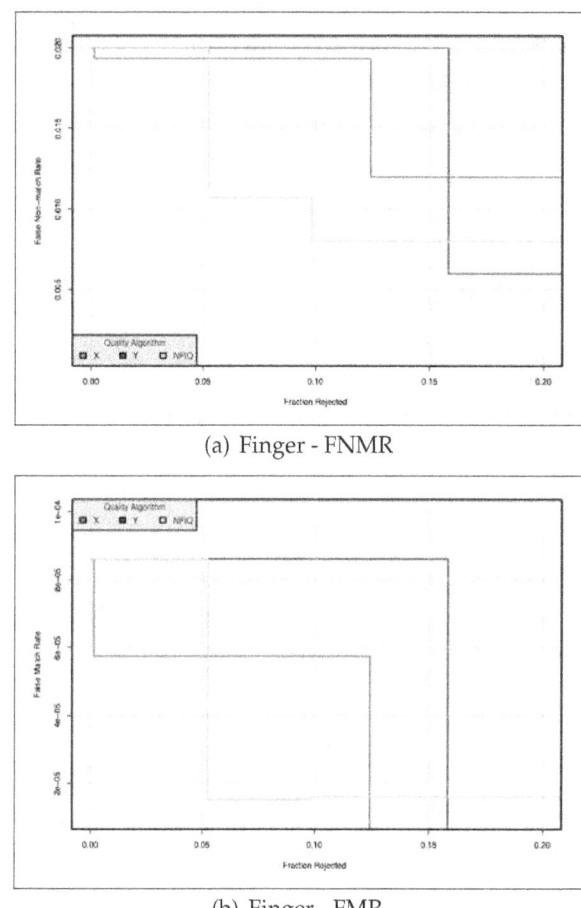

(a) Finger - FNMR

(b) Finger - FMR

Figure 5.2: Error vs. reject performance for three fingerprint quality methods. Figures (a) and (b) show reduction in FNMR and FMR at a fixed threshold as up to 20 % of the low-quality samples are rejected. The similarity scores come from a commercial matcher.

If the quality values are perfectly correlated with the genuine scores, then when we set t to give an overall FNMR of x and then reject proportion x with the lowest qualities. A recomputation of FNMR should be zero. Thus, a good quality metric correctly labels those samples that cause low genuine scores as poor quality. For a good quality algorithm, FNMR should decrease quickly with the fraction rejected. The results of applying this analysis are shown in Figure 5.2. Note that the curves for each of the three fingerprint quality algorithms trend in the correct direction, but that even after rejection of 20 %, the FNMR value has fallen only by about a half from its starting point. Rejection of 20 % is probably not an operational possibility unless an immediate reacquisition can yield better quality values for those persons. Yoshida, using the same approach, reported similar figures [22]. Note, however, that for NFIQ, the improvement is achieved after rejection of just 5%. In verification applications such as access control, the prior probability of an impostor transaction is low, and thus the overall error rate is governed by false non-matchers. In such circumstances, correct detection of samples likely to be falsely rejected should drive the design of BQAMs.

5.3 Generalization to Multiple Matchers

It is a common contention that the efficacy of a quality algorithm is necessarily tied to a particular matcher. We observe that this one-matcher case is commonplace and useful in a limited fashion and should therefore be subject to evaluation. However, we also observe that it is possible for a quality algorithm to be capable of generalizing across *all* (or a class of) matchers, and this too should be evaluated.

Generality to multiple matchers can be thought of as an interoperability issue: can supplier A's quality measure be used with supplier B's matcher? Such a capability will exist to the extent that pathological samples do present problems to both A and B's matching algorithms. However, the desirable property of generality exposes another problem: we cannot expect performance to be predicted absolutely because there are good and bad matching systems. A system here includes all of the needed image analysis and comparison tasks. Rather we assert that a quality algorithm intended to predict performance generally need only be capable of giving a relative or rank ordering i.e., low-quality samples should give lower performance than high-quality samples.

The plots of Figure 5.3 quantify this generalization for the NFIQ system using the error vs. reject curves of section 5.2. Figure 5.3(a) includes five traces, one for each of five verification algorithms. The vertical spread of the traces indicates some disparity in how well NFIQ predicts the performance of the five matchers. A perfectly general BQAM would produce no spread.

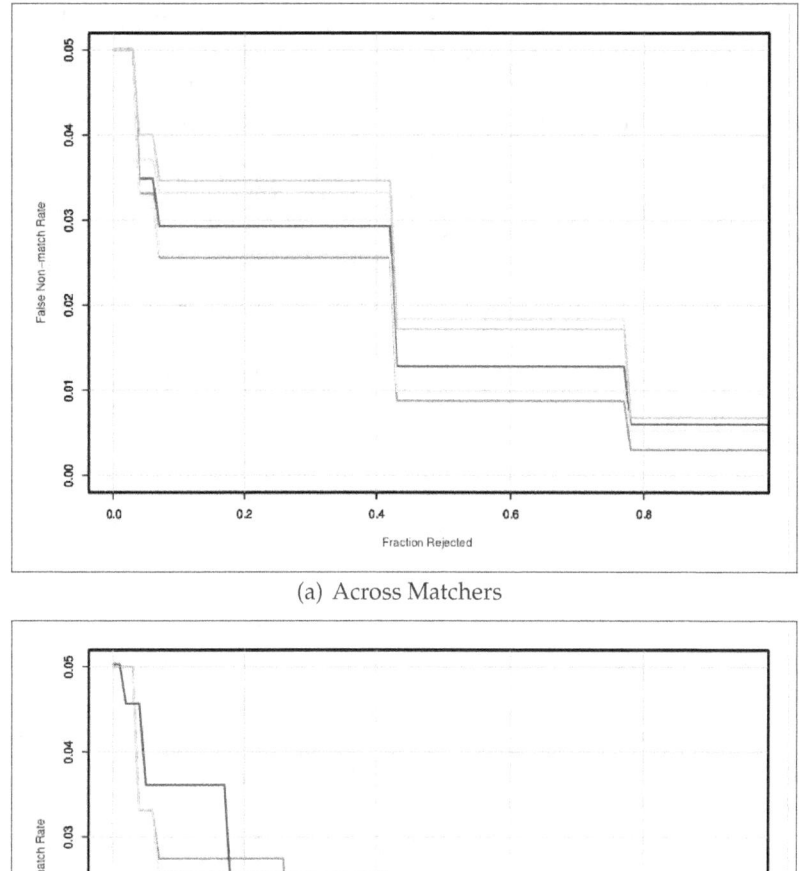

(a) Across Matchers

(b) Across Datasets

Figure 5.3: Error vs. reject characteristics showing how NFIQ generalizes across (a) five verification algorithms, and (b) three operational data sets. The steps in (a) occur at the same rejection values because the matchers were run on a common database.

Table 5.1: KS statistics for quality levels of three quality algorithms

Quality Algorithm	$Q=1$	$Q=2$	$Q=3$	$Q=4$
Quality Algorithm 1	0.649	0.970	0.988	0.993
Quality Algorithm 2	0.959	0.995	0.996	0.997
Quality Algorithm 3	0.918	0.981	0.994	0.997

5.4 Measuring Separation of Genuine and Impostor Distributions

We can evaluate quality algorithms on their ability to predict how far a genuine score will lie from its impostor distribution. This means instead of evaluating a quality algorithm solely based on its FNMR (i.e., genuine score distribution) prediction performance, we can augment the evaluation by including a measure of FMR because correct identification of an enrolled user depends both on correctly finding the match and on rejecting the non-matches. Note also that a quality algorithm could invoke a matcher to compare the input sample with some internal background samples to compute sample mean and standard deviation.

The plots of Figure 5.4 show, respectively, the genuine and impostor distributions for adjusted NFIQ values, 1, 3, and 5. The overlapping of genuine and impostor distributions for the poorest NFIQ means higher recognition errors for that NFIQ level, and vice versa; the almost complete separation of the two distribution for the best-quality samples indicates lower recognition error. NFIQ was trained to specifically exhibit this behavior.

We consider the Kolmogorov Smirnov (KS) statistic. The KS test is non-parametric, distribution-free, and simple. The KS statistic is simply the maximum absolute difference between the two distributions' cumulative distributions functions. For better-quality samples, a larger KS test statistic (i.e., higher separation between genuine and impostor distribution) is expected. Each row of Table 5.1 shows KS statistics for one of the three quality algorithms that we tested. KS statistics for each quality levels $u = 1, \ldots, 5$ are computed by first computing the genuine (i.e., $\{s_{ii} : (i,i) \in R(u)\}$) and impostor (i.e., $\{s_{ij} : (i,j) \in R(u), i \neq j\}$) empirical cumulative distributions, where $R(u) = \{(i,j) : H(q_i^{(1)}, q_j^{(2)}) = u\}$. Thereafter the largest absolute difference between the genuine and impostor distributions of quality u is measured and plotted. (Note that to keep quality algorithm providers anonymous, we only reported KS statistics of the lowest four quality levels.)

5.5 Data to be Used for Testing

A quality measurement algorithm could be evaluated using data specifically collected with deliberate defects. For example, quality could be degraded by misfocusing the camera.

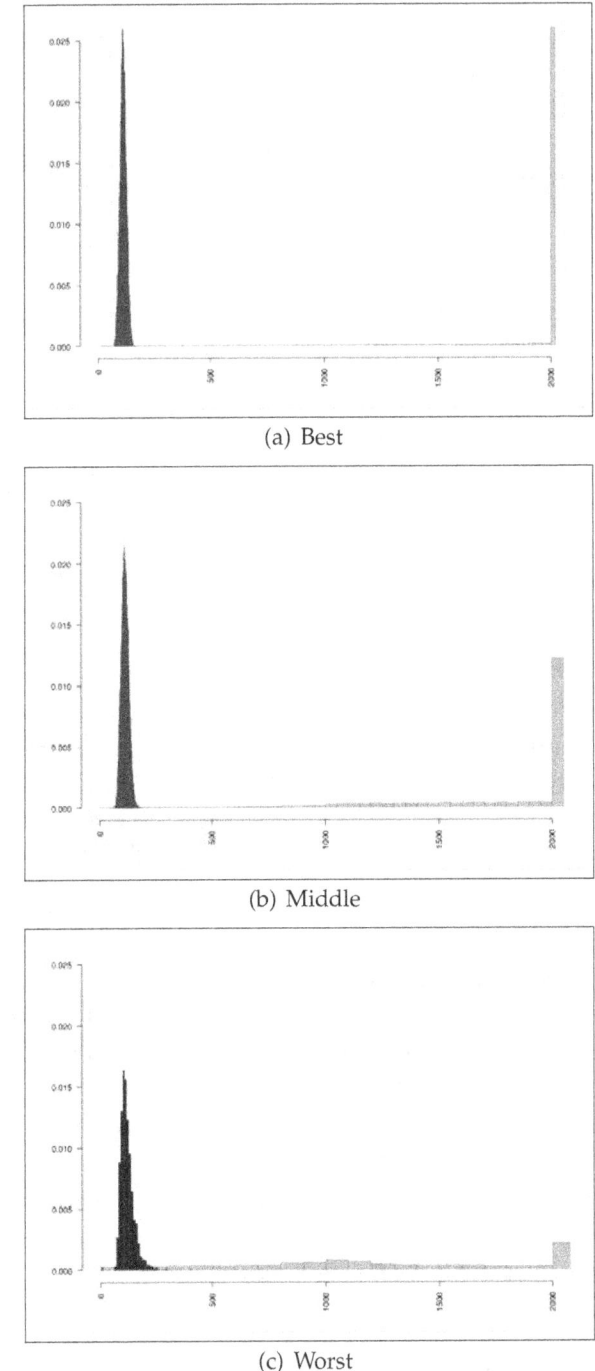

(a) Best

(b) Middle

(c) Worst

Figure 5.4: There is a higher degree of separation between the genuine and impostor distribution for better-quality samples as measured by NFIQ.

Such data have several notable uses: development of a quality measurement algorithm, teaching best practice by counterexample, and assessing the performance of a product intended to test the conformity of an image or signal to an underlying standard.[3] However, we argue that this type of data should not be used for evaluation for four reasons. First, such data is by definition laboratory data and therefore would lack application-specific operational relevance. Second, by applying certain kinds of degradation to the images, the evaluator is making assumptions about the performance sensitivities of matching algorithms. For example, if the chin is cropped from a face image, then this may be immaterial to a face recognition algorithm. Third, it would be difficult or impossible to collect samples that express all possible combinations of quality defects and particularly with their natural frequency and to their natural degree. Finally, the laboratory data may not ordinarily be available in large quantities.

Instead the use of operationally representative data, i.e., samples harvested during real-world usage or from a relevant scenario test [23] should be considered. By definition, this has the advantage of having relevance to the operation. We showed examples of such data in Section 5.2.

To illustrate the importance of using an aggregated corpus for evaluation, we use the Color FERET database [24]. The frontal *fa* and *fb* images from each of 852 subjects were used at full, half, and quarter resolutions. These are input to a quality algorithm and a matching algorithm from the same supplier. The reduction in image size forcibly induces the reductions in both quality and match scores evident in Figure 5.5. Note, however, that for any one of the three point clouds in Figure 5.5(a), there is large variation in score in relation to quality - a trend that is not improved by plotting $M(s)$ instead (Figure 5.5(b)). This reflects the difficulty of the face quality problem.

The final graph, Figure 5.5(c), shows the error versus reject performance for each of the image sizes separately and for the aggregate data set. This latter curve, in grey, is lower than the others. This demonstrates the value of using composite sets for evaluation purposes. Also worthy of note is that the error versus reject performance at any of the three sizes is superior to that in Fig. 5.5, which uses the same algorithm on a more uniform dataset. Those images are about the same size as the half-size FERET images but are more consistently posed (i.e., frontal), sized, compressed, and all subjects do not wear eyeglasses. The suggestion, then, is that the more homogenous the corpus, the more difficult it is for a quality algorithm to predict variation in similarity scores. We should emphasize that the algorithm was provided to NIST without any claim of efficacy or recommended domain of use.

---

[3]For example, the ISO/IEC 19794-5 Face Recognition Interchange Format standard puts quantitative limits on the amount of quality-related degradation as blur, non-frontal pose, and the number of grey levels.

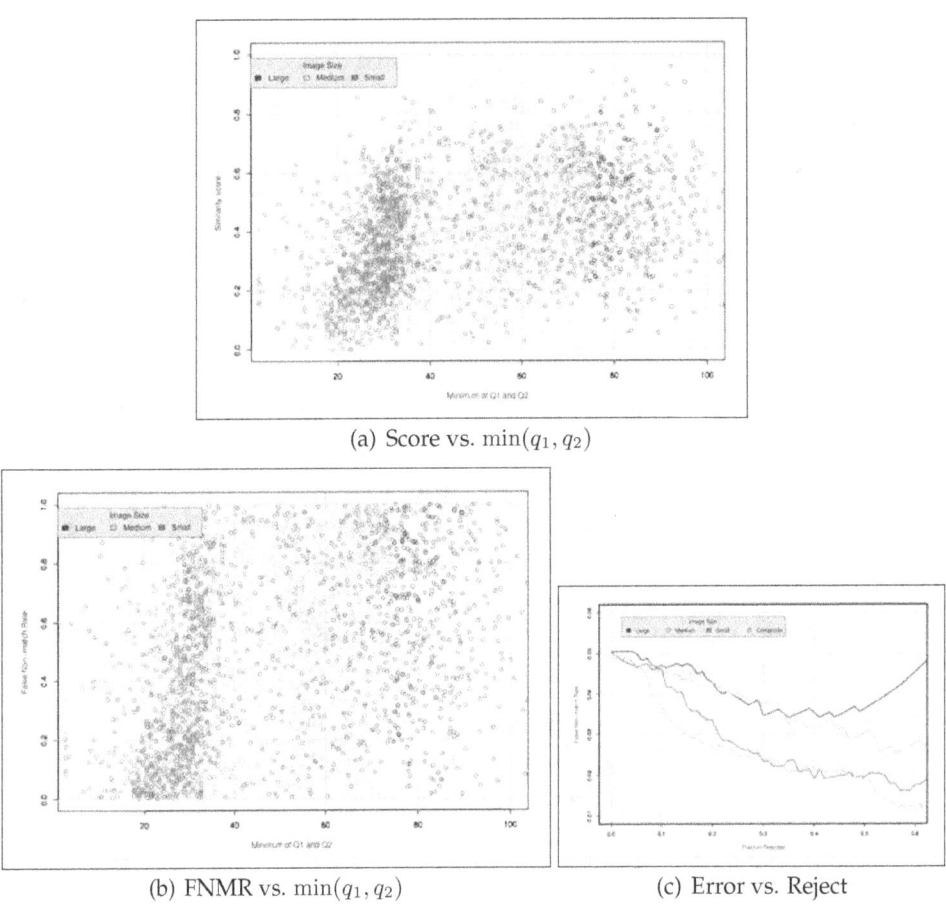

(a) Score vs. $\min(q_1, q_2)$

(b) FNMR vs. $\min(q_1, q_2)$

(c) Error vs. Reject

Figure 5.5: Scatter plots of scores and FNMR values versus quality, and the error vs. reject curves, for a face quality metric applied to a face database composed of images at full (blue), half (green), and quarter size (red).

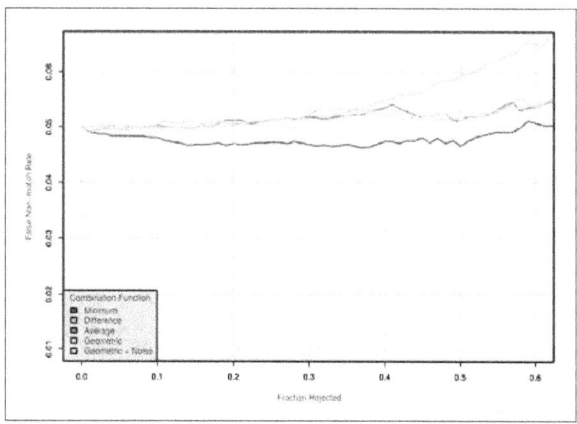

Figure 5.6: Error vs. reject characteristic of same algorithm as in Figure 5.5 but using data with less quality variation. The plots correspond to different quality combination function H(.) as discussed in Section 4.4.

Chapter 6

Quality Reference Data Sets

This chapter recommends a procedure to annotate the samples of a reference corpus with quality values. Quality-annotated corpus could be used for quality algorithm development. quality calibration, and conformity of quality scores to a standard.

6.1 Construction of a Reference Data Set

The strategy is to assign values that are directly related to the results of matching those samples. This is achieved by taking the similarity scores from $K \geq 1$ matching algorithms, classifying them, and, in the case of $K > 1$, taking a consensus. The result is a reference set useful to quality algorithm developers. It would be of use for tuning of an operational quality algorithm, when the matcher and kind of data are known.

The input to the procedure is a representative sample database. The output is an annotation of each sample with a scalar quality target. The method presumes the availability of a representative matching algorithm, which will be used to compare samples to produce both genuine and impostor similarity scores. It is therefore implied that two or more samples per person are available.

6.1.1 Data

Data gathered in a target operational application would be most realistic. Contemporary matchers perform extremely well on most images, and it is therefore necessary to preferentially stack the reference set with samples that are naturally problematic to the matcher. For example, for a reference fingerprint data set to span the quality spectrum, it should be, to the degree possible, balanced in terms of finger position (right/left index/thumb/middle), finger impression (roll/plain/flat), sex, age, and capture device. Lack of data often renders it difficult to create such a balanced dataset.

6.1.2 Target Quality Assignment

We seek to assign a performance-based quality score to each image in a reference dataset. We ensure that the quality values are representative of performance by associating the image with similarity scores as follows. Consider a biometric corpus containing two samples, $d_i^{(1)}$ and $d_i^{(2)}$, for each of N individuals, $i = 1, \ldots, N$. The first samples represent enrollment samples, and the second samples represent those for authentication. The following procedure assigns quality values $q_i^{(1)}$ and $q_i^{(2)}$ to all images in the corpus.

1. For each matching algorithm V_k, $k = 1, .., K$ of K available algorithms:

   For each person i:

   (a) Compare the first and second samples using the k-th matcher to produce genuine score. Repeating equation 4.3:
   $$s_{ii}^{(k)} = V_k(d_i^{(1)}, d_i^{(2)}) \tag{6.1}$$
   where V_k is the k-th matching algorithm for all available $k = 1, ..., K$ matching algorithms.

   (b) Compare the first sample from person i with the second sample from all $j = 1, \ldots, N$ and $i \neq j$ other persons. The result is $J = N - 1$ impostor scores,
   $$s_{ij}^{(k)} = V_k(d_i^{(1)}, d_j^{(2)}) \tag{6.2}$$
   (This is essentially equation 4.4.)

   (c) Insert i into set $\mathcal{T}$ if its genuine score is larger than all its impostor scores, i.e., $s_{ii}^{(k)} > s_{ij}^{(k)} \; \forall j$. This is a rank 1 condition.

   (d) For the first sample of each person $d_i^{(1)}$, compute the sample mean and standard deviation of its J associated impostor scores
   $$m_i = \frac{\sum_{j=1}^{J} s_{ij}^{(k)}}{J-1} \tag{6.3}$$
   $$\sigma_i = \sqrt{\frac{\sum_{j=1}^{J} \left(s_{ij}^{(k)} - m_i\right)^2}{(J-1)^{-1}}} \tag{6.4}$$

(e) Normalize the genuine score from eq. 6.1 using the impostor statistics

$$z_i = (s_{ii} - m_i)/\sigma_i \tag{6.5}$$

Once all normalized similarity scores have been computed:

(a) Compute two empirical cumulative distribution functions: One for the top-ranked genuine scores of set $\mathcal{T}$

$$C(z) = \frac{|\{z_i : i \in \mathcal{T}, z_i \leq z\}|}{|\{z_i : i \in \mathcal{T}, z_i \leq \infty\}|} \tag{6.6}$$

and another for those not in that set.

$$W(z) = \frac{|\{z_i : i \notin \mathcal{T}, z_i \leq z\}|}{|\{z_i : i \notin \mathcal{T}, z_i \leq \infty\}|} \tag{6.7}$$

These cumulative distribution functions are plotted in Figure 6.1 for live-scan images of the right-index fingers of 6000 individuals and scores of a commercial fingerprint matcher. These were produced in a U.S. government test using sequestered operational data.

(b) Bin normalized match score range into K bins based on quantiles of the normalized match score distribution. One strategy, for $K = 5$, is shown in Table 6.1 in which F^{-1} is the quantile function, and $F^{-1}(0)$ and $F^{-1}(1)$ denote the empirical minima and maxima, respectively. If $W^{-1}(1) \geq C^{-1}(0.25)$ an appropriate quartile of $C(z)$ must be selected.

(c) Sample d_i is assigned target quality q_i corresponding to the bin of its normalized match score z_i from eq. (6.5).

(d) The procedure is repeated for sample $d_i^{(2)}$ by swapping indices 1 and 2 in equations 6.1 and 6.2. Since one sample will have an impostor distribution different from another, two different samples of the same subject may have different normalized match scores and therefore different quality values.

2. Samples with identical quality assignments from *all* V matchers become members of the Quality Reference Dataset. Those without unanimity are discarded.

This procedure has been used to form NFIQ training and compliance set [8], only with different bin boundaries. These were set by manual inspection to give useful categorization of the normalized match score statistic.

Table 6.1: Binning normalized match score

Category	Description	Range of normalized match score
1	poor	$\{z_i : -\infty \leq z_i < C^{-1}(0)\}$
2	fair	$\{z_i : C^{-1}(0) \leq z_i < W^{-1}(1)\}$
3	good	$\{z_i : W^{-1}(1) \leq z_i < C^{-1}(0.25)\}$
4	very good	$\{z_i : C^{-1}(0.25) \leq z_i < C^{-1}(0.75)\}$
5	excellent	$\{z_i : C^{-1}(0.75) \leq z_i\}$

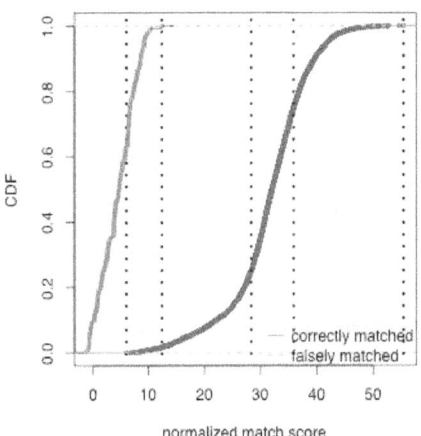

Figure 6.1: Empirical cumulative distribution functions for the top-ranked genuine scores and for the impostor scores. The vertical lines are one possible way of binning normalized match scores. Samples are assigned quality numbers corresponding to the bin of their normalized match score.

Chapter 7

Fingerprint Image Quality

A fingerprint is a pattern of friction ridges on the surface of a fingertip. A good-quality fingerprint has distinguishable patterns and features that allow the extraction of features, which are useful for subsequent matching of fingerprint pairs. A minutia-based automatic fingerprint matching algorithm uses features that compare local ridge characteristics (minutia) of two fingerprints and produces a real-valued similarity score.

Several factors affect the quality of fingerprint images: user's skin condition, improper finger placement, scanner limitation or imperfection, impurities on the scanner surface, and others. Using the terminology of Section 1.1, the cause of these imperfections can be classified in four groups: *i) impairments in the source of biometric (character):* like scars, blisters, skin conditions such as wet or dry, age, occupation, etc.; *ii) user behavior:* such as improper finger placement, e.g., rotating finger or placing only tip of a finger which cause capturing insufficient area of finger image; *iii) imaging:* for example, low contrast, distortion, sampling error, insufficient dynamic range, etc.; and *iv) environment:* such as temperature, humidity, or unclean platen.

7.1 NIST Fingerprint Image Quality (NFIQ)

NFIQ is a fingerprint quality measurement tool; it is implemented as open-source software conformant to the ISO/IEC 9899:1999 C specification, and is used today in U.S. government and commercial deployments. Its key innovation is to produce a quality value from a fingerprint image that is directly predictive of expected matching performance, and has been designed to be matcher-independent. Definition of quality as prediction of performance first introduced by NFIQ has been widely adopted by industry and the research community, and consequently the international biometric sample quality (ISO/IEC 29794) is currently under development.

NFIQ first measures appropriate image fidelity characteristics. These quality components are then fused using a three-layer feed-forward nonlinear perceptron model so that the overall quality score is prediction of recognition errors likely to be realized when the sample is matched. NFIQ extracts minutia, assigns a quality value to each minutia point, and measures orientation field, pixel intensity, and directional map to compute the following local and global features: number of foreground blocks, number of minutia, number of minutia that have quality value better than certain thresholds, percentage of foreground blocks of excellent, good, fair, and poor quality. A neural network was trained to classify the computed feature vectors into five levels 1 through 5 where NFIQ = 1 is the best-quality and NFIQ = 5 is the lowest quality. The neural network was trained on how far a sample's genuine score would lie from its impostor distribution. Fig. 7.1 shows that the highest recognition performance is achieved for the best quality samples (NFIQ=1), and samples with lowest quality (NFIQ=5) have the lowest performance. In Section 5.4, we noted that an effective BQAM would exhibit better separation of genuine and impostor distributions for better-quality samples. The plots of Fig. 5.4 (in Section 5.4) show, respectively, the genuine and impostor distributions for NFIQ values 1(excellent quality), 3(average quality), and 5(poor quality). The overlapping of genuine and impostor for the poorest NFIQ (i.e., NFIQ=5) means higher recognition errors for that NFIQ level, while the almost complete separation of the two distributions for the best-quality samples (i.e., NFIQ=1) indicates lower recognition error. Source code for NFIQ algorithm is publicly available and can be downloaded from [7].

7.2 Recommendations for NFIQ Summarization

This section, as promised in Section 2.2, recommends procedures for NFIQ summarization. The motivation for NFIQ summarization is to monitor quality variation over time, across different acquisition settings and/or application.

In an operation where fingerprint images are collected and their NFIQ values are computed, the overall quality of the collection is given by:

$$\tilde{Q} = 102.75 - 2.75p_1 - 5.37p_2 - 14.38p_3 - 42.25p_4 - 102.75p_5 \qquad (7.1)$$

where p_i is the proportion of the fingerprints with quality value $i = 1\ldots 5$. The weights were determined using the method of Appendix A, and they reflect the likelihood that an observed false non-match involved a fingerprint of quality i. The terms of equation 7.1 indicate that the errors are dominated by images with NFIQ values 4 and 5, and this implies that a plain averaging of observed values is not an appropriate summary. Thus users of NFIQ should not use the mean or median of a set of quality values as a summary statistic. Equation 7.1 produces a NFIQ summary on the range $[0, 100]$. This is achieved by a transformation of a simpler linear quantity (see the development in Appendix A). It is used here to allow standardized range of biometric sample quality values; mainly

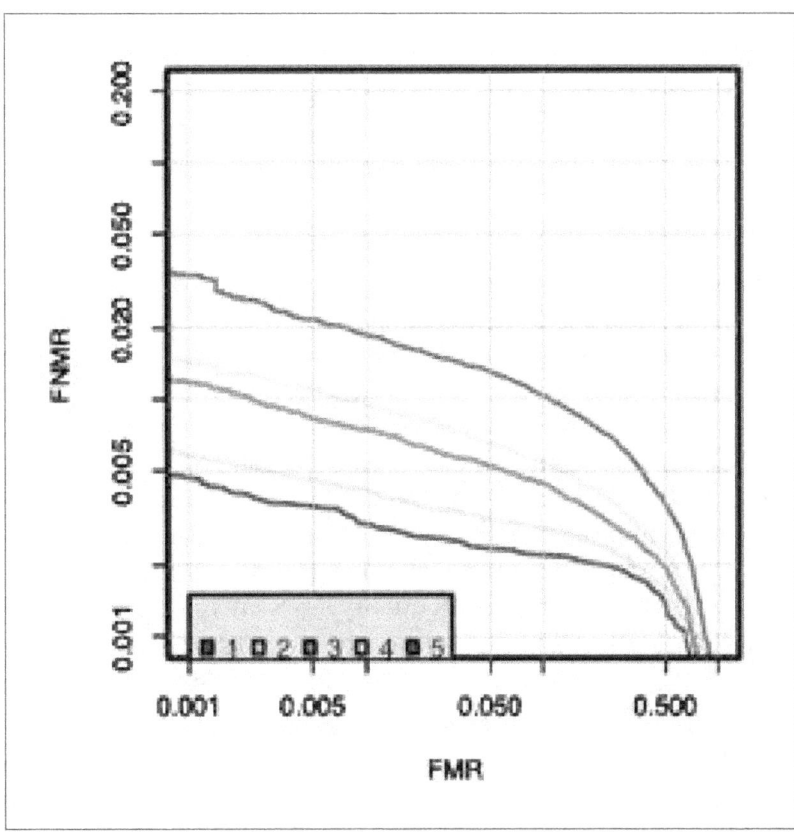

Figure 7.1: Quality-ranked detection error trade-off characteristics. Five traces correspond to five NFIQ levels. Fingerprint images with NFIQ=1 (highest quality) cause lower recognition error than images with NFIQ=5 (lowest quality).

in keeping with the ISO/IEC 19784-1 BioAPI [17] requirement for single sample quality values on $[0, 100]$. Equation 7.1 provides a Best Practice estimate for the NFIQ algorithm for those verification applications in which the specific matchers and operating thresholds are unknown. Discussion of dependance on matcher algorithm and operating threshold follows.

> **Dependence on Matching Algorithm** Weights in equation 7.1 are consensus estimates. That is, they were estimated using the observed false non-match rates from a set of leading commercial matching algorithms. The result is that the weights are not exactly the weights that would be used for any one algorithm, or for a specified set of algorithms. NIST regards the NFIQ weights above as Best Practice estimates to be used unless other details about the application are known.

> **Dependence on Operating Threshold** Weights in equation 7.1 are estimates of the observed false non-match rates computed at some fixed threshold. The result is that these weights are most accurate for that particular threshold and not as accurate for biometric systems operating at other thresholds. Figure 7.2 shows the variation of these weights computed at three different thresholds. It appears that weights for NFIQ values of 1 and 2 are quite robust to a wide range of thresholds, but weight for NFIQ value 5 varies with threshold. Table 7.1 shows recommendation for NFIQ summarization at several operation threshold. NIST regards these recommendations as Best Practice estimates, and these should be used unless other details about the application are known. Thus, in verification applications, where operating threshold is fixed at τ, users of NFIQ fingerprint quality assessment algorithm should either use the weights computed at threshold closest to τ (as shown in Table 7.1) or follow the procedure in Appendix A to establish dedicated weights.

In verification applications, where a specific set of one or more matching algorithms or operating thresholds are known and available, users of NFIQ fingerprint quality assessment algorithm should follow the procedure in Appendix A to establish dedicated weights.

Table 7.1: Recommendation for NFIQ summarization at different operating thresholds

False Match Rate	Recommendation for NFIQ summarization
0.01	$101.91 - 1.91 p_1 - 3.97 p_2 - 10.24 p_3 - 34.03 p_4 - 101.91 p_5$
0.001	$102.75 - 2.75 p_1 - 5.37 p_2 - 14.38 p_3 - 42.25 p_4 - 102.75 p_5$
0.0001	$105.41 - 5.41 p_1 - 9.15 p_2 - 23.82 p_3 - 55.81 p_4 - 105.41 p_5$

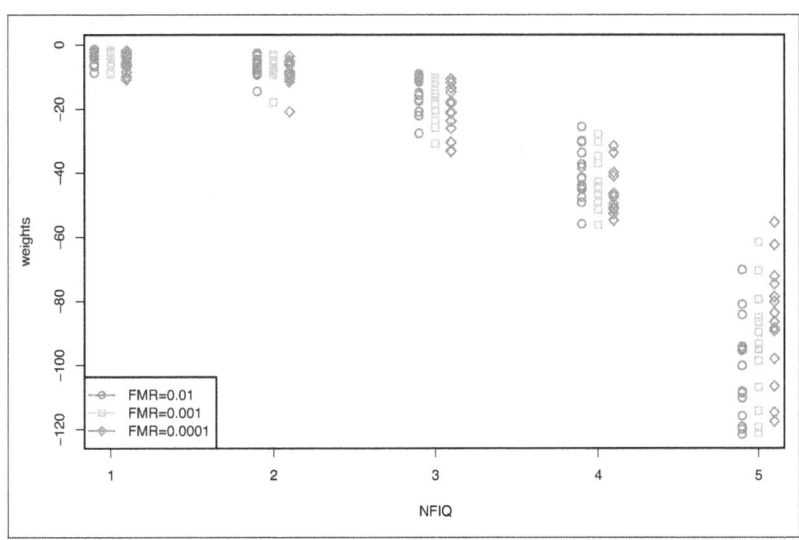

Figure 7.2: Dependance of NFIQ weights on operating threshold. Weights for NFIQ values 1 and 2 are quite robust to variation of the computing threshold. Thresholds are set at overall false-match-rates of 0.01, 0.001, and 0.0001. Each point corresponds to the NFIQ weight estimated using similarity scores of a commercial matching algorithm on large operational fingerprint datasets. NFIQ weights in Table 7.1 are means of six matching algorithms with the highest performance.

Closing

Biometric quality measurement is an operationally important and difficult problem that is nevertheless massively under-researched in comparison to the primary feature extraction and pattern recognition tasks. In this paper, we enumerated the ways in which it is useful to compute a quality value from a sample. In all cases, the ultimate intention is to improve matching performance. We asserted, therefore, that quality algorithms should be developed to explicitly target matching error rates, and not human perceptions of sample quality. To this end, we defined a procedure for the annotation of a reference sample set with target quality values. We gave several means for assessing the efficacy of quality algorithms. We reviewed the existing practice, cautioned against the use of detection error trade-off characteristics as the primary metrics, and instead advanced boxplots and error vs. reject curves as preferable. We suggest that algorithm designers should target false non-match rate as the primary performance indicator.

In conclusion, we posit that quality summarization as a predictor of recognition performance is a difficult problem, and we encourage the academic community to consider the problem and extend the quantitative methods of this paper in advancing their work.

The benefit of measuring and reporting of biometric sample quality is to improve performance of biometric systems by improving the integrity of biometric databases and enabling quality-directed processing in particular when utilizing multiple biometrics. Such processing, enhancements result in increasing probability of detection and track accuracy while decreasing probability of false alarms.

Appendix A

Determination of Quality Weights

This section advances a procedure for assigning weights to the output values of a BQAM. We assume quality values are quantized into L levels so that (without loss of generality) $q = 1 \ldots L$, where $q = 1$ and $q = L$ indicate lowest and highest quality values, respectively. This is the case with NFIQ for which $L = 5$ and other commercial BQAMs for which $L = 8$ and $L = 10$. The strategy is to assign weights u_q that are directly related to the error rate observed for samples of quality q.

Suppose some enterprise collects fingerprints and measures the quality of each. If the number of prints collected over some interval in an operational situation is n and this is composed of n_q prints of quality q, then we could compute the mean quality across all n samples. However, arithmetic mean is not the preferred method of summarizing quality scores because all samples, regardless of their quality values, are given the same weight. If instead the expected utility of a fingerprint of quality q is $u_q = U(q)$, then a better summary statement of quality is

$$\bar{q} = \frac{\sum_{q=1}^{L} u_q n_q}{\sum_{q=1}^{L} n_q} \tag{A.1}$$

If the utility u_q is actually an estimate of the false reject rate for samples of quality q of a reference fingerprint verification system operating at some reasonable threshold, then $\bar{q}$ will be an estimate of the expected error rate. We proceed by introducing a procedure to compute utility u_q for different levels of a BQAM such that the summarized quality value is an estimate of the expected error rate.

Consider a biometric corpus contains $2N$ pairs of images from N persons. The first sample represents an enrollment sample, and the second represents the authentication sample. The samples have integer qualities $q_j^{(1)}$ and $q_j^{(2)}$ for $j = 1, \ldots, N$. Applying V matching algorithms to the samples, we get

▷ N genuine similarity scores, $s_{jj}^{(v)}$, and

▷ up to $N(N-1)$ impostor scores, $s_{jk}^{(v)}$ with $j \neq k$

where $v = 1, \ldots, V$ and $V \geq 1$.

1. For all matching algorithms v and quality values q, compute $\text{FNMR}^v(\tau, i)$ of authentication samples of quality i with enrollment samples of quality better than or equal to i at operating threshold τ using genuine scores of matching algorithm v. Note that we assumed higher quality values indicate better quality. For BQAMs which low values indicate good quality (for example, NFIQ), $q_j^{(1)} \leq i$, $q_j^{(2)} = i$ should replace $q_j^{(1)} \geq i$, $q_j^{(2)} = i$ in the computation of $\text{FNMR}^v(\tau, i)$ below.

   for $(v = 1, \ldots, V)$
      for $(i = 1, \ldots, L)$

$$\text{FNMR}^v(\tau, i) = \frac{\left|\left\{s_{jj}^{(v)} : s_{jj} \leq \tau, \; q_j^{(1)} \geq i, \; q_j^{(2)} = i\right\}\right|}{\left|\left\{s_{jj}^{(v)} : s_{jj} \leq \infty, \; q_j^{(1)} \geq i, \; q_j^{(2)} = i\right\}\right|}$$

      end
   end

which results in the following array

$$\begin{pmatrix} \text{FNMR}^1(\tau, 1) & \text{FNMR}^2(\tau, 1) & \ldots & \text{FNMR}^V(\tau, 1) \\ \text{FNMR}^1(\tau, 2) & \text{FNMR}^2(\tau, 2) & \ldots & \text{FNMR}^V(\tau, 2) \\ \ldots & \ldots & \ldots & \ldots \\ \text{FNMR}^1(\tau, L) & \text{FNMR}^2(\tau, L) & \ldots & \text{FNMR}^V(\tau, L) \end{pmatrix}$$

2. compute weight u_i

$$u_i = \frac{\sum_{v=1}^{V} \text{FNMR}^v(\tau, i)}{\sum_{q=1}^{L} \sum_{v=1}^{V} \text{FNMR}^v(\tau, q)}$$

Thus the aggregated quality across an enterprise is

$$Q = \sum_{i=1}^{L} u_i p_i \qquad (A.2)$$

where u_i are estimated posterior probabilities above. As probabilities, these values will not be on a range familiar to users. For example, the NFIQ summary is

$$Q = 0.016 p_1 + 0.032 p_2 + 0.086 p_3 + 0.252 p_4 + 0.613 p_5 \qquad (A.3)$$

such that if all samples were of NFIQ = 1 (i.e., the best quality), the result would be $Q = u_1 = 0.016$. Similarly the worst case is when all samples in the enterprise are of NFIQ = 5, which results in $Q = u_5 = 0.613$. Thus this formulation would result in NFIQ summaries on the range $[u_1, u_5]$, which is $[0.016, 0.613]$. Users should regard equation A.3 as a measure of expected overall FNMR. However, this document recommends transformation from $[u_1, u_5]$ to the more familiar BioAPI [17] range $[0, 100]$ which has 0 as the lowest quality and 100 as the best. This can be accomplished by:

1. Either relating the quality summary number Q (i.e., expected error rate) back to the native quality range by using the inverse of the utility function:

$$\tilde{Q} = U^{-1}(Q) = U^{-1}\left(\sum_{i=1}^{L} u_i p_i\right) \tag{A.4}$$

    where U^{-1} is a function approximation (e.g., piece-wise linear interpolation) of pairs (i, u_i);

2. Or by mapping (e.g. linear mapping) $[u_1, u_5]$ to $[0, 100]$. Thus, NFIQ summaries mapped to $[0, 100]$ are given by

$$\tilde{Q} = \frac{100 u_5}{u_5 - u_1} - \sum_{i=1}^{5} \frac{100 u_i}{u_5 - u_1} p_i \tag{A.5}$$

which forms equation 7.1 in this document. Table 7.1 shows recommendation for NFIQ summarization at different thresholds where utility $u_i\ i = 1\ldots 5$ (i.e., five levels of NFIQ) is computed at different operating thresholds, and linearly mapped to $[0, 100]$ using equation A.5.

Bibliography

[1] T. Ko and R. Krishnan, "Monitoring and reporting of fingerprint image quality and match accuracy for a large user application," in *Proceedings of the 33rd Applied Image Pattern Recognition Workshop.* IEEE Computer Society, 2004, pp. 159–164.

[2] *Procedings of the NIST Biometric Quality Workshop.* NIST, March 2006, http://www.itl.nist.gov/iad/894.03/quality/workshop/presentations.html.

[3] D. Benini et al., *ISO/IEC 29794-1 Biometric Quality Framework Standard*, 1st ed., JTC1 / SC37 / Working Group 3, Jan 2006, http://isotc.iso.org/isotcportal.

[4] Y. Chen, S. Dass, and A. Jain, "Fingerprint quality indices for predicting authentication performance," in *Procedings of the Audio- and Video-based Biometric Person Authentication (AVBPA)*, July 2005, pp. 160–170.

[5] E. Tabassi, *Fingerprint Image Quality, NFIQ*, NISTIR 7151, National Institute of Standards and Technology, 2004.

[6] E. Tabassi, "A novel approach to fingerprint image quality" in *IEEE International Conference on Image Processing ICIP-05*, Genoa, Italy, September 2005.

[7] *NIST Fingerprint Image Quality (NFIQ) source code.* NIST, August 2004, http://www.itl.nist.gov/iad/894.03/nigos/nbis.html.

[8] E. Tabassi, *NFIQ Compliance Test, NISTIR 7300*, National Institute of Standards and Technology, http://fingerprint.nist.gov/NFIQ, 2006.

[9] F. Alonso-Fernandez, J. Fierrez-Aguilar, and J. Ortega-Garcia, "A review of schemes for fingerprint image quality computation," in *COST 275 - Biometrics-based recognition of people over the Internet*, October 2005.

[10] J. Ortega-Garcia, J. Fierrez-Aguilar, D. Simon, J. Gonzalez, M. Faundez-Zanuy, V. Espinosa, A. Satue, I. Hernaez, J.-J. Igarza, C. Vivaracho, D. Escudero, and Q.-I. Moro, "Mcyt baseline corpus: a bimodal biometric database," *Proceedings of the IEE Conference on VISP*, vol. 150, no. 6, pp. 395–401, December 2003.

[11] E. Lim, X. Jiang, and W. Yau, "Fingerprint quality and validity analysis," in *Proceedings of the IEEE Conference on Image Processing*, vol. 1, September 2002, pp. 469–472.

[12] Bioscrypt Inc., *Systems and Methods with Identify Verification by Comparison and Interpretation of Skin Patterns such as Fingerprints*, June 1999, http://www.bioscrypt.com.

[13] L. M. Wein and M. Baveja, "Using fingerprint image quality to improve the identification performance of the u.s. visit program," in *Proceedings of the National Academy of Sciences*, 2005, www.pnas.org/cgi/doi/10.1073/pnas.0407496102.

[14] J. Fierrez-Aguilar, J. Ortega-Garcia, J. Gonzalez-Rodriguez, and J. Bigun, "Discriminative multimodal biometric authentication based on quality measures," *Pattern Recognition*, vol. 38, no. 5, pp. 777–779, May 2005.

[15] E. Tabassi, G. W. Quinn, and P. Grother, "When to Fuse Two Biometrics," in *IEEE Computer Society Conference on Computer Vision and Pattern Recognition CVPR-06*, New York, June 2006, Biometric Workshop.

[16] C. Tilton et al., *The BioAPI Specification*, American National Standards Institute, Inc., 2002.

[17] ISO/IEC JTC1 / SC37 / Working Group 2, *ISO/IEC 19784-1 Information Technology - Biometric Application Programming Interface - Part 1: BioAPI*, 2006, http://isotc.iso.org/isotcportal.

[18] ISO/IEC JTC1 / SC37 / Working Group 3, "ISO/IEC 19794 Biometric Data Interchange Formats," 2005, http://isotc.iso.org/isotcportal.

[19] A. J. Mansfield, *ISO/IEC 19795-1 Biometric Performance Testing and Reporting: Principles and Framework*, FDIS ed., JTC1 / SC37 / Working Group 5, August 2005, http://isotc.iso.org/isotcportal.

[20] J. Fierrez-Aguilar, L. Muñoz-Serrano, F. Alonso-Fernandez, and J. Ortega-Garcia, "On the effects of image quality degradation on minutiae and ridge-based automatic fingerprint recognition," in *IEEE International Carnahan Conference on Security Technology*, October 2005.

[21] D. Simon-Zorita, J. Ortega-Garcia, J. Fierrez-Aguilar, and J. Gonzalez-Rodriguez, "Image quality and position variability assessment in minutiae-based fingerprint verification," *IEE Proceedings on Vision, Image and Signal Processing*, vol. 150, no. 6, pp. 395–401, December 2003, special Issue on Biometrics on the Internet.

[22] A. Yoshida and M. Hara, "Fingerprint image quality metrics that guarantees matching accuracy," in *Procedings of NIST Biometric Quality Workshop*. NEC Corp., March 2006, http://www.itl.nist.gov/iad/894.03/quality/workshop/presentations.html.

[23] M. Thieme, *ISO/IEC 19795-2 Biometric Performance Testing and Reporting: Scenario Testing*, cd2 ed., JTC1 / SC37 / Working Group 5, August 2005, http://isotc.iso.org/isotcportal.

[24] *The Color FERET Face Database*, National Institute of Standards and Technology, http://www.nist.gov/humanid/feret, March 2002.

[25] A. Martin, G. R. Doddington, T. Kamm, M. Ordowski, and M. A. Przybocki "The DET curve in assessment of detection task performance," in *Proceedings of Eurospeech*, pp. 1895–1898, Rhodes, Greece 1997.

[26] A. J. Mansfield, and J. L. Wayman, *Best practices in testing and reporting performance of biometric devices*, National Physics Laboratory Report CMSC 14/02, August 2002, http://www.cesg.gov.uk/site/ast/biometrics/media/BestPractice.pdf.

[27] J. M. Chambers, W. .S. Cleveland, B. Kleiner, and P. .A. Tukey, *Graphical Methods for Data Analysis*, Wadsworth and Brooks/Cole, pp. 62, 1983.

[28] R. M. McCabe et al., *Data Format for the Interchange of Fingerprint, Facial, and Other Biometric Information*, ANSI/NIST, 2007.

[29] Grother, P. et al.: MINEX: Performance and Interoperability of the INCITS 378 Fingerprint Template. National Institute of Standards and Technology. NISTIR 7296 (2005) http://fingerprint.nist.gov/minex04.